Educating the Whole Child
in
Catholic Education

Yanny C. Salom

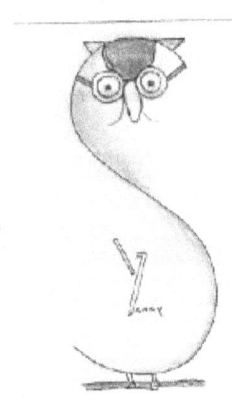

No part of this book may be reproduced or transmitted in any manner whatsoever without written permission, except for the use of quotations in books, or articles, or reviews for inclusion in a newspaper, magazine, broadcast, or website. For more information, contact owlcreations2024@gmail.com

Front Cover Design by Corrie Leas
Edited by Larissa Rosa and Andrea Salom
ISBN: 979-8-9913932-4-9

Copyright © 2025 by Yanny C. Salom
All rights reserved.

Library of Congress Control Number: 2025903473
Published in the United States of America
First printing, March 2025

Printed in Seattle, Washington.

This book is dedicated to the Holy Name of Jesus[1]

"There is salvation in no one else, for there is no other name under heaven given among mortals by which we must be saved." Acts 4:12

---

[1] *If you want to learn more about the devotion to the Holy Name of Jesus, go to* https://www.holyname.ie/saint-bernardine-of-siena-and-the-power-of-the-name-of-jesus/ This will take you to a website that has already compiled a list of prayers, reflections, and resources for your use to assist in the creation of an age-appropriate lesson plan for your students.

Contents

Acknowledgements	vii
About the Author	ix
Educating the Whole Child Statement	x
Prologue	xi
Chapter 1: Recognizing Students' Identity	1
Chapter 2: Connecting Our Faith	5
Chapter 3: STEM Education and Its Value on the Conservation Movement	9
STEM Integration "Sheltering and Monitoring Bluebirds"	11
STEM Integration "Saving Florida Sea Turtles in their Race for Survival"	23
Chapter 4: We Are Better Together	33
Roller Coasters	34
STEM Books	43
Constellations	50
The Free-Market Project	64
Chapter 5: Center for STEM Education – University of Notre Dame	67
Modeling Science Phenomena	68
Assessments in STEM Integrations	72
Embracing the Uniqueness of the Human Person	82
STEM Blog – Notre Dame Center for STEM Education	85
Chapter 6: The Value of STEM in Food Science	87
STEM Lab "The Science Behind Baking Cookies"	92
Chapter 7: Creating and Coordinating an Astronomical Society	95
Chapter 8: Teaching our Faith	101
Chapter 9: An Integrated Approach for Foreign Language Acquisition	107
Miscellaneous	111
Words of Advice – Class of 2022-2023	112

HoFAS Article from AFA Falcon Newsletter	114
Butterfly Garden	115
Article: Our New STEM Lab (August 2022)	116
Blessed Carlo Acutis STEM Laboratory and Plaque Statement	118
Works Cited	119

Acknowledgements

My heartfelt gratitude to:

My loving parents, Jairo and Mildred, for the inheritance of my faith and the gift of Catholic education.

My dear husband, Juan Carlos, his support took many forms; drilling holes, cutting wood, turning eggs in the incubator, cleaning the hydroponics garden, being a speaker, serving as a judge in the science and engineering fair, helping in the water tower competition, attending the stargazing nights, maintaining the butterfly garden, driving me to Lakeland for the Northeast Florida Regional Science and Engineering Fair and the list goes on.

My sweet daughter, Andrea, for our endless conversations about Catholic matters and for showing me with her own experience outside the Catholic environment how to educate children in secular environments. Her faith is my inspiration and ongoing joy.

Father Timothy Cusick, for always modeling the teacher-minister role and for finding time out of his busy schedule to visit my students in the classroom. They had questions and doubts that he welcomed and responded to help them build understanding of their faith. My gratitude for Father Tim's effort to build this bridge cannot be overstated.

The Cascone Family Foundation, their support to Catholic education is admirable and greatly appreciated. My educational dreams for students came true thanks to their commitment to Catholic education, their generosity, and their willingness to enable our children's growth through a meaningful educational experience.

School leaders, their open-door policy came alongside their willingness to be open to new ideas and perspectives. They always encouraged and supported me to implement new educational initiatives and teaching methods.

Church leaders, for integrating Catholic practices and traditions into the school, and for joining efforts with the school community to foster faith, which is the essential pillar of Catholic education.

The members of the Center for STEM Education at the University of Notre Dame, a network of professionals that are a support and a resource to one another.

Mrs. Rosemary Nowotny, for encouraging me to pursue a teaching career in science. She inspired and helped me recognize my God–given gift.

Mrs. Corrie Leas, her support and availability to collaborate with projects truly enhanced all we did as members of the middle school cluster. She continued collaborating with me after her retirement.

Mrs. Meghan Chakhtoura and Mr. Michael Kavanagh, their commitment to intentional interdisciplinary connections were of great service in the development of the STEM program and culture.

Mrs. Larissa Rosa, for co-coordinating the astronomical society from the beginning of this initiative. Her perseverance to find speakers, coordinate our stargazing nights, and bake galactic snacks were a gift that we all enjoyed year after year.

Our wonderful middle school cluster teachers, without them the STEM integrations would not have been possible. Their professionalism and sense of collaboration were the answers to my prayers.

Our elementary school teachers, for always opening their minds, hearts, and classrooms to the joy of interdisciplinary activities.

Our school parents, for always collaborating with God's Providence. Their support and appreciation were undeniable and a source of joy. They truly fueled me with the energy I needed to continue building the program.

Mrs. Juliette Gredenhag, one of my English professors at Florida Community College and, later, one of those encouraging parents; her support and kind words represent an example of the importance of empowering second language learners to find a purpose as they assimilate to new cultures.

Mr. Mike Wolf and Mrs. Mary Wolf, for donating their time, expertise, and school birdhouse. Mrs. Amy Lacy, for continuing the family legacy by caring for the birds and helping with scientific observations.

Mrs. Brandi Salomone and Mrs. Lynn Rosario, for maintaining the butterfly garden during the final years before my retirement. An honorable mention to our alumna, Amanda Waldrep, for her sustained support to the maintenance of the butterfly garden.

My students, their enthusiasm and love for learning always lifted my spirits. They were the reason for the endless hours spent creating learning opportunities. I would do it all over again for them. We also switched roles many times and they taught me about technology and English pronunciation. I truly received more than what I gave them as their teacher.

The many individuals, authors, and institutions that I have mentioned and referenced throughout the book.

About the Author

Yanny Salom is a civil engineer who is passionate about structural design and Catholic education. She worked as an engineer for five years and spent five years teaching physics at the College of Engineering in Maracaibo, Venezuela, where she graduated. Yanny earned her Bachelor of Science in Civil Engineering and her Master of Science in Education and Applied Science from the Universidad del Zulia in her hometown. Once in the United States of America, she started her career as a teacher-minister in the Diocese of St. Augustine. Yanny began teaching Spanish, then added science laboratory class. She spent fifteen years teaching middle school science, including nine of those years teaching seventh grade religion. She coordinated engineering competitions, science fairs, after school programs, summer camps, interactions with scientists, science family nights, activities, and clubs that brought science alive for all.

As a Notre Dame STEM Trustey fellow, she was able to expand the scope of her educational programs infusing new initiatives in education. The Air Force Association Falcon Chapter honored her with the 2015 Teacher of the Year and the 2017 Member of the Year award for her energy and enthusiasm to stimulate students through project-based learning with real world problems and applications in STEM and Aerospace Education. She established the Holy Family Astronomical Society to encourage students to look beyond our planet by reaching into space with awe and reverence to God's creation. She was also awarded by the Catholic Climate Covenant for her achievement in environmental sustainability and stewardship through the conservation education program that she developed and facilitated to her students. Her scientific knowledge, professional background, and faith–based scientific approach provided academic rigor and a strong ethical and moral component to her lessons, empowering children to utilize their potential and be agents of change in the world.

She is enthusiastic about educating the whole child within an interdisciplinary approach that puts faith at the center to lead souls to Heaven.

Educating the Whole Child within Catholic education goes beyond academic matters and raising good citizens, as Catholic faith is seamlessly integrated to see God in all things and to lead souls to Heaven. It calls for seeing the intersection of faith and reason to arrive at logical conclusions considering the ethical and moral implications of our decisions. Reaching eternity in Heaven is the cornerstone of Catholic education, as we help students:

- Understand that 'freedom' implies the pursuit of what is right, what is real, and what is lovely
- Develop a firm and habitual disposition towards the good
- Acknowledge that God has gifted each person with inherent worth and dignity
- Recognize that the language of our body, as a God-given gift, must speak the truth
- Engage in a shared life within society working towards the common good

The task of forming a worldview requires students to be engaged, motivated, supported and challenged in a safe environment, fostering an education that responds to the needs of the whole person.

Prologue

These days of technological advances have brought great opportunities and substantial challenges. We have all heard about artificial intelligence, but my heart continues to choose the value of human relationships. Technology will continue providing great resources, but it will never be a substitute for human interactions. God created us with free will and rationality. Our rationality can certainly help us create technology to learn something, since all the intelligent answers we can get from a new technology were first discovered by human intelligence. The capacity to think, as well as the capacity to choose, are inherent to a being with a rational soul, the human person. In that sense, we can develop great technology, but it is limited because we cannot provide it a spirit or free will. Whatever the machine does, is the product of a set of established neural connections and directives that will make a decision based on what a person has directed. This is why disconnecting a machine does not have moral implications, because it is not a human person. In the same order of ideas, this is also why the worth of a human act is so valuable, unique, and unreproducible. "Freedom makes a man a moral subject. When he acts deliberately, man is, so to speak, the father of his acts." (CCC #1749). We should assert the good of technology but should also reflect on how technology that is not used with regulation and self-control may isolate humans from God and other human beings.

The pages of this book contain some best practices and reflections with an emphasis on my sense of gratitude to the amazing group of people who were part of this educational journey. Establishing relationships increases the success of the educational programs and enhances the school atmosphere. Furthermore, significant learning occurs when we develop remarkable relationships with our students. Being able to lean on Christ through the relationships we establish, is an important skill in Catholic education. Relationships are not perfect, but the Holy Spirit is always showing us the path of compassion, kindness, and empathy to maintain our human connections. The Word of God is our guide and inspiration, as we were created for loving relationships with God and others. Through deepening our relationship with God, we can improve our relationship with others and become a sign of God's love for others.

I cannot end this section without providing my insight about middle school students. I encountered a group of young people who wanted to be loved and accepted. Loving them unconditionally and reaching them exactly where they were without judgment gave them a great sense of belonging. Our classroom culture is something that I will always miss. They were my emotional learners, with the ups and downs of adolescence, but the constants were their love, their desire to help me with any project, and their willingness to teach me something in return. During my last years in the classroom, they were instrumental in the creation of STEM projects for students in the elementary levels. There is so much goodness in our children, and for those days when their choices were not aligned with God's plan for them, they received my advice without judgement and the certainty of a fresh start after discussing the consequences of their actions. Those are precisely the times when we need to embrace them and assure them that they belong in God's Kingdom and are a precious part of the Mystical Body of Christ.

Educating the whole child has deep implications in Catholic education. God has made each individual in his image and likeness as we exist in this body-soul composite that makes us a human person. Our commitment goes beyond academic matters as we want to help students to know who they really are and to honor the uniqueness of each person. Unfortunately, mass media and social media platforms, show a distorted image of reality and have changed young people's understanding of freedom. Our youth need to fix their eyes on all the goodness God has created and have the perseverance to be part of the goodness we see in the world. The Bible tells us the wonderful story of God's plan for salvation, and he will never stop helping humanity. He became one of us through Jesus Christ and continues to inspire us with holy people so we can increase our understanding of the truth about who we are. Educating the whole child means living out the gospel for our children and showing them through our actions that freedom implies doing God's will. Reaching eternity in Heaven must be the cornerstone of Catholic education. St. Augustine perfectly captured that mission in his famous quote "You have made us for yourself, O Lord, and our hearts are restless until they rest in you." (Chadwick 3).

God is a God of patience. A long time ago, I thought that building and designing school facilities as an engineer was God's plan for me. With full awareness that subject knowledge was necessary, but not enough to teaching, I continued my preparation to fulfill God's will in my vocational journey. Once in the United States, I learned English and became certified to teach Spanish K-12 and Science 5-9. It was then that I fully understood that God's plan for me was to cooperate with him to shape the young living stones of his Church, the students, the recipients of the Teachings and Traditions of the Church. St. Mother Teresa said in an interview, "the work is only the expression of the love we have for God. We have to pour our love on someone. And the people are the means of expressing our love." (Towey 4). This quote represents my gratitude to God for the gift of the vocation of teaching and for my dear students who were God's instruments to express my love.

It is with great joy and a grateful heart that I share these pages with you, the educator. Being an educator is not exclusive to professionally trained individuals, as parents are the primary educators of their children. You will find as an essential practice "the integration of our faith" with other academic disciplines with an emphasis on our role as agents of change. This is my humble intent of compiling some of my best practices and reflections with the support of some great books that I used to expand my knowledge and stay updated on new educational initiatives. You will also find the Bible and the Catechism of the Catholic Church being cited often as they were my best tools to bring the Word of God to my students. Forming them in academic matters was important, but shaping their consciences to fit God's plan for eternity was essential. As a last thought in this prologue, I just want to reaffirm with certainty that the future is bright! I know that because God allowed me to take a quick look at it in my classroom.

Chapter 1

<u>Recognizing Students' Dignity</u>

"But you have upheld me because of my integrity and set me in your presence forever."

Psalm 41:12

When I look back at nearly two decades as a Catholic school teacher, the most powerful takeaway is the value of living out the gospel to educate, advocate, pray, and bring Jesus to the students' lives. I probably brought this practice from the Franciscan Catholic school I attended as a child. In a Franciscan setting, living the gospel is the daily invitation of our patron saint, St. Francis of Assisi. In that sense, Christ's call into the teaching ministry implies the important task of leading students into a life-transforming relationship with him. To be able to bring Jesus to the children we teach, we need to start by encountering Jesus in them. In other words, we need to see our students as children of God created in his image since this is the only way to recognize their dignity. God has gifted each person with worth and dignity, which requires an education that responds to the needs of the whole child.

Teaching comes with the daily struggles of grading, curricula, behavior, meetings, and many other rigors of the profession. We cannot get lost in the process of challenging the advanced and advancing the challenged because our job goes beyond that duty. We must nurture their minds while we feed their souls and show them how their identity is in Christ. I have always found inspiration in the life of St. Mother Teresa of Calcutta. She strived to see Jesus in others and rescuing the dignity of the human person was a common factor in everything she did. Her life, insight, and actions continue to be a constant reminder of the value of respect, empathy, and compassion to love well. This led me to make an intentional effort of judging students' actions rather than judging their character, which is without a doubt a better way to assess and provide feedback about behaviors.

In my personal experience, the best strategy was building a relationship with students. This approach allowed me to reach students academically in a more effective way, manage behavior, understand certain behaviors or attitudes, foster a safe classroom, and reach students at the spiritual level. Knowledge of students' likes and dislikes facilitated conversations where they could easily engage. The development of a genuine regard for their interests was important too, but sharing my stories emphasized my humanity and made me more approachable to them. The more human you are, the more they realize that struggles and making mistakes are part of the path to success. Finally, and most importantly, have a sense of humor because learning can happen when having fun.

To be consistent with recognizing in students the dignity with which God created them, discipline must take place with authority, but also with profound love and respect. Stressing the power of a "fresh start" is a meaningful and authentic way to provide space for redemption. We tend to believe that we know the whole story, but the story is not complete until we hear what our students have to say. In addition, this is a great opportunity to teach our students to be accountable for their actions and to own a plan of action to practice virtues instead of vices.

One practice that I found helpful and meaningful when students made poor choices was asking students to choose a virtue they have failed to practice. If a student's behavior exemplified a lack of the virtues of prudence and charity, then they practiced those virtues for seven days reflecting on the process by filling out the "Virtue Chart" you will find on the next page. The basis behind this approach is the concept of the human body-soul composite: the body makes visible the invisible reality via actions and body language. Following this concept, practice makes perfect whether that is the practice of virtue or vice. I emphasize the three theological virtues; faith, hope, and charity as well as the four cardinal virtues; prudence, justice, fortitude and temperance. However, other values that are indicators of good character were included as they fit.

I found a very similar practice in the book *Theology of the Body for Teens*, the difference being that they practice the virtue for ten days and students can choose a friend, as an accountability partner, to help them grow in the chosen virtue. Seven days or ten days will not be enough to master it, but it is a good way to create that expectation in students that will settle in their heart and eventually permeate in their decisions. One more advantage I found with this strategy was that we focused on how to help children to become a better version of the person God has created them to be rather than attaching adjectives to the students who have performed the poor behavior. This is truly a Christian approach where the person's dignity is preserved. We live in a polarized world where people are quick to establish and share opinions about others. We need to train the upcoming generations of students to show more compassion and empathy. Protecting the reputation of others, which is an aspect of the dignity of the human person, is essential.

Practicing Virtues

Virtue Chart

Chosen Virtue:

Day	What did you accomplish?	What were your challenges or temptations?	What is your plan to accomplish your goal?
1			
2			
3			
4			
5			
6			
7			

Discuss with your teacher or the people involved in the situation your experience after practicing the chosen virtue.

What was helpful to practice the virtue?

How are you going to continue integrating this practice in your life?

What are some effective ways you can help family members, or the school community, to be better on modeling the chosen virtue?

Educating the Whole Child in Catholic Education

Chapter 2

Connecting Our Faith

"Faith and reason are like two wings on which the human spirit rises to the contemplation of truth."

St. John Paul II ~ Encyclical Letter "Fides Et Ratio"

As teacher-ministers, we must connect our faith to any subject we teach. Educating the whole child calls for seeing the intersection of faith and reason to arrive to logical conclusions, and even more important, to show reverence to God, the Creator of all things. Middle school science and STEM gave me the opportunity to make that connection. The task of bringing faith and reason within a single worldview came with challenges, but I found strength and inspiration from the fact that Jesus continued to teach despite the doubts of others and the rejections he encountered.

As a science teacher, wonder is a basic scientific skill that has the advantage of strengthening our relationship with creation and with reality. An authentic and true Catholic education teaches students to see reality as it is, both physically and spiritually, since we exist in an integrated universe that God has created. Catholicism is for people who think, as Catholics employ both faith and reason in the day-to-day realities of life and ministry. This idea engages with this section about faith and reason, but even more importantly, it invites us to think critically to arrive at a logical conclusion.

My role as an educator went beyond the content and practices of the subject of science, given the integration of our Catholic beliefs and values as essential components of being a teacher-minister. One of the life science chapters that brings up the religious and scientific perspectives is "evolution by natural selection". When these two perspectives seem to be in contradiction, the key to finding common ground is the use of reason to know that integration was built into the natural law. Below are articles to direct the class discussion towards the conclusion that God created the world freely and wisely to better reveal his glory through the intelligent evolutionary process that takes place within his providential will.

In response to the topic of the theory of evolution from a Catholic perspective, the articles found within the Thomistic evolution website can provide a starting point for discussions in the middle school classroom[2]:

---

[2] To learn more about evolution from a Catholic perspective, visit https://thomisticevolution.org.
If you wish to learn more, the Thomistic Evolution website offers a High School Curriculum.

From the essay *The Fittingness of Evolutionary Creation*
"For the Catholic theological tradition, the answer to the purpose-of-creation question is clear: God chose to create because he wanted to manifest and to communicate his glory." (Austriaco 1).

From the essay *How does God create through Evolution?*
"The changing creature can only be changing because it was created as such by an unchanging Creator." (Austriaco 2).

From the essay *God's Knowledge and Love in Creation*
"God understood exactly what he was doing when he created each and every aspect of reality, and everything that exists does so because he loves it, including each and every one of us." (Davenport 2).

From the essay *The Nature of Creation*
"Biology and ecology both reveal a particularly breathtaking beauty, harmony, and order of things. We can see that there is an intelligible plan for the world, and we can see what the plan is to some extent (at least to the extent of knowing something of the general order of nature), even though we cannot discover God's plan for each particular thing in detail." (Brent 2).

There are a number of science concepts that are part of the foundation of evolution, like inheritance. When teaching inheritance, we must also discuss critical concepts on biotechnology. The main categories of biotechnology include artificial selection, genetic engineering, and cloning (Dispezio et al. 163). Studying biotechnology goes beyond science and touches on issues that may be morally wrong. Cloning has been used to copy segments of DNA for genetic engineering as well as to make copies of DNA samples to confirm the identity of a person who has committed a crime. Scientists have cloned mammals, which raises serious concerns and implications. Class discussions must help students recognize that understanding the science of genetics and the Teachings of the Church will allow them to make informed decisions on when biotechnology can be used. Catholicism opposes human cloning since it is a depersonalized way of bringing a new human being into the world. God has revealed himself to us in the person of Jesus, who was begotten, not made.

Since cloning overlaps with issues of stem cells, this provides an opportunity to explain that Catholic teaching is opposed to the destruction of human life, and this is the reason it opposes embryonic stem cell research. However, stem cell research that comes from the umbilical cord and adult tissue do not constitute a moral dilemma. As teachers, we know that students respond better to an assertive command. For example, asking to walk instead of saying "do not run" will create a more positive interaction and outcome with students. The same applies to our faith and that is why promoting unity and focusing on the aspects we can agree upon is vital. We must explain why a practice is morally wrong, but we cannot end the discussion without offering alternatives that fit the Church's Teachings.

This way we provide a powerful tool for students to understand our faith in a more effective and positive way, as we help develop their confidence in defending their faith. The following excerpt from the Catechism of the Catholic Church represents an excellent starting point for discussion: "Research or experimentation on the human being cannot legitimate acts that are in themselves contrary to the dignity of persons and to the moral law. The subjects' potential consent does not justify such acts." (CCC # 2295). The United States Conference of Catholic Bishops has relevant information on bioethics that you will find useful and meaningful in the following two:

1. Bioethics Materials - Pro Life Activities | USCCB
   https://www.usccb.org/prolife/bioethics-materials
2. Resources in Catholic Bioethics — The Catholic Bioethics Center (ncbcenter.org)
   https://www.ncbcenter.org/bioethics-resources

Before we start learning about inheritance, we have to explain how cells divide for sexual reproduction. Meiosis is the type of cell division that produces haploid sex cells known as eggs and sperm cells. Two good practices during this part of the program are: sending a letter to parents about the vocabulary and content that is being taught; and using the *Theology of the Body for Teens* program to bring our Catholic identity to the lesson. This program for seventh grade students utilizes components of the *Theology of the Body for Teens* curriculum designed by Ascension Press and other resources, and it is based on St. John Paul II's brilliant teaching. The program explores how God graciously created the human person with inherent dignity and value, and how this affects relationships and moral choices.

The cases explained above give us an obvious opportunity to connect with our faith, but we can connect with our faith even in the least obvious cases. Another example was a simple investigation we performed to practice measurements and the International System of Units in which the research question asked: Does the rind of an orange affect its behavior when placed in water? This experiment gave students the opportunity to practice scientific skills related to the scientific method as well as the concepts of density and buoyancy, but the outcome of the experiment provided space for a relevant analogy that connected with our Catholic faith. The orange floated on the surface before removing the rind, but it sank after removing it. The science behind the outcome is related to density. The rind of the orange has air pockets that reduce the density of the orange to a value that is less than the density of water (objects less dense than water will float). Removing the rind increases the density of the orange, making it sink. The orange with the rind represents those who wear the Armor of God because they can stay afloat when temptations come into their lives. When we do not wear the Armor of God, we start moving away from the truth that we can find in the Word of God. Ephesians 6: 10-17 provides biblical support for this lesson.

In the Catholic school I attended as a student, we had the advantage of having Franciscan priests and sisters as teachers. It made the integration of faith and science a natural and daily approach with interesting conversations in the classroom. One priest and philosophy teacher used to ask questions to make us think and defend Church's Teachings, which is an important skillset to develop as missionary disciples. Inviting priests to the classroom was a fruitful way to see the science subject through an integrated perspective. Students were really interested in what priests had to say about topics we covered in science and religion classes, and they also formulated interesting questions that bridged the gap between science and faith. The value of having priests as speakers is that they represent an authority that facilitates bringing Christ into the students' lives. In other words, this practice helped me bring students into a closer relationship with Jesus and his Church regarding morality, Scriptures, Catholic Social Teaching, and the teachings found in the *Catechism of the Catholic Church*.

St. John Paul II shared a meaningful metaphor that provides insight about the relationship between faith and reason. In his Encyclical Letter *Fides Et Ratio,* he stated that "faith and reason are like two wings on which the human spirit rises to the contemplation of truth." Blessed José Gregorio Hernández Cisneros was a physician, a scientist, and a professor, but above all, he was a man of faith who lived the gospel through the Corporal and Spiritual Works of Mercy. His faith intersected beautifully with his knowledge of medicine to help the sick, the suffering, and the poor. This remarkable lay Franciscan is known as the 'doctor of the poor.' He performed his profession in a scientific manner without disregarding moral laws, and in a manner that would not conflict with his faith. In the same vein, Albert the Great found in the natural phenomena and in creation a way to glorify God, so science and theology were intercepted to show the perfection of God, the Creator. "Albert loved science and philosophy because he loved God" (Vost 45). Let's ask St. Albert, the patron saint of scientists, to help us grow in faith and reason, and to always see God in all things.

Chapter 3

STEM Education and Its Value on the Conservation Movement

> "This responsibility for God's earth means that human beings, endowed with intelligence, must respect the laws of nature and the delicate equilibria existing between the creatures of this world."
>
> Encyclical Letter Laudato Si, 83 ~ Pope Francis

All human activities have short and long-term consequences for ecosystems. Also, all the ethical considerations that those consequences raise become an opportunity for our students to consider the impact of their decisions and their solutions. Highly effective STEM education would help students recognize the multiplicity of solutions while weighing the implications for people and the environment. STEM prioritizes real-world environmental issues and project-based learning, helping students to appreciate the planet on which we live. Pope Francis's encyclical letter, *Laudato Si'*, has within it a call to action to ecological protection and stewardship of our home. Both the encyclical letter and STEM programs have common ties that, if integrated, will result in a deeper understanding of complex global environmental dilemmas and the solutions to resolve them. A STEM approach gives teachers the opportunity to teach ethics as our classrooms become spaces with a meaningful civic purpose that allow us to educate the whole child about the natural and material world around them. Educating the whole child entails more than raising good citizens but raising stewards of all God has created as a requirement of our faith. We are called to enter into a relationship with the creation where God manifests as the Creator. St. Francis of Assisi invites us to live in communion with each other and with nature without excluding anyone.

Some of the student-made projects on environment are listed below:

- Building and monitoring bluebird houses to alleviate the impact of human deforestation.
- Creating a habitat for monarch butterflies so they can find a safe place to raise their young on our campus.
- Creating awareness about the race for survival of sea turtles in Florida, engineering filtration systems to reduce industrial pollutants discharged to the ocean, and writing letters to Florida senators to encourage them to support efforts to ensure the survival of sea turtles swimming in Florida's waters.
- Solving environmental problems for their Science and Engineering Fair projects. Their ideas ranged from charging cell phones by converting energy from living plants into electrical energy to engineering an eco-friendly bioplastic to use in utensils that makes them biodegradable. In botany, a student proposed the use of polymers from diapers to engineer a soil that could maintain humidity to mitigate water loss in dry climate.

- Working on endangered species projects where students wrote reports on threatened or endangered species that included the description of the species, its habitat, evidence of why it is endangered, and conservation efforts to protect the animal. Students also adjusted to scale a realistic 3D model of the endangered species for an educational board game with the goal of raising awareness about the struggles of those species.
- Symbolically adopting endangered species where students wrote paragraphs to convince the school community to adopt their chosen species through the World Wildlife Foundation (WWF). To make this even more community oriented, the school's Astronomical Society and Nature Club designed ballots for school wide voting after a presentation of the persuasive paragraphs.
- Addressing another local problem, which is the rising mortality rate of our Florida manatees. We know that human activities like boating are some of the main reasons for the struggle of this herbivorous mammal. After completing our unit on sound waves, our eighth-grade classes worked on a STEM analysis on the effect of boats on manatees' population and provided solutions to reduce the risk of manatees being killed or hurt by boats. Boats seem to be the leading cause of death of this marine animal and the unit on sound waves gave students the scientific knowledge to consider effective alerting devices to deliver solutions.
- Creating Public Service Announcements (PSA) of different environmental problems to create awareness as students were positioned as helpful humans. The goal was teaching others how to change their behaviors to solve those environmental problems. These PSAs were broadcasted through the school's social media platforms.
- Addressing the Plastic Problem[3]: millions of living organisms are killed by plastic every year and we can make a difference by reducing the amount of plastic waste we generate. This gave students the opportunity to use the Engineering Design Process (EDP) to provide a solution related to our habitual consumption of plastic.
- Creating videos of products that use sustainable energy and highlight the benefits of using that alternative energy.
- Teaching students farming skills through the embryology project (baby chicks), butterfly project and hydroponics garden: it teaches students how to care for animals, life cycles and where food comes from as they develop a greater appreciation for nature and for food.
- Incorporating STEM Books on environment (see Chapter 4 for further details).

These conservation initiatives were designed to foster collaboration, critical thinking skills, and communication while working through the Engineering Design Process (EDP) with evidence-based data that allow students to generate designs and see the value of multiple solutions. Students were living out the Catholic Social Teaching "Care for God's Creation" as they reflected on their roles as

---

[3] This STEM integration was one of the projects I chose to earn my National Geographic Educator Certification. To learn more about this free professional development program, visit https://www.nationalgeographic.org/society/education-resources/profesional-development/

stewards of the environment. They modeled St. Francis of Assisi, the patron saint of Ecology, while they learned the practical value of scientific knowledge and considered the world of engineering as a possible career path.

Following, you will find details about the STEM integration on bluebirds and the STEM integration on Florida sea turtles. The whole-child approach to education must include appreciation of the natural world as we help children to position themselves as stewards of all God has created. It includes the preservation of habitats as discussed through the STEM integration on sheltering and monitoring bluebirds. It also encompasses ensuring clean water as discussed through the STEM integration on Florida sea turtles.

STEM Integration "Sheltering and Monitoring Bluebirds"

Highly effective STEM education helps students recognize the value of multiple solutions while considering the implication on people and the environment. Bluebirds nest in cavities, but human deforestation destroys habitats. The learning objectives of this STEM integration are:

1. Using the Engineering Design Process (EDP) to design birdhouses for bluebirds.
2. Monitoring bluebird activity and nesting habits.

To provide a clear picture of the STEM integration, a summary of the curriculum alignment has been added, the STEM integration experience, the Next Generation Science Standards (NGSS), and a description of the project. The first step to design a plan is the creation of a chart that shows the curriculum alignment. This practice allowed cluster teachers to know which disciplines were involved and to intentionally align the STEM integration objectives with the diocesan objectives. Science, math, religion, and engineering practices were seamlessly combined to solve the problem of these fascinating perching birds. Since many disciplines are involved, the STEM Integration Experience chart describes:

1. The Activity and its Description
2. Content and Practice Goals
3. Assessment Choice

The STEM Integration Experience chart also shows how to:

1. Contextualize the problem with data on population and scientific articles about bluebirds.
2. Create career connection, which in this case is through a speaker to discuss the role and job as an environmental engineer. We also had a speaker who was an expert on local birds.
3. Delimit the problem by determining constraints that are based on birds' characteristics.
4. Using the EDP to develop possible solutions.
5. Communicate solutions.

The next chart corresponds to the NGSS which includes:

1. Science and Engineering Practices
2. Discipline Core Ideas
3. Crosscutting concepts

The project description includes:

1. Problem
2. Constraints
3. EDP: ask, imagine, plan, create, and improve
4. Communicate
5. Catholic identity (Catholic Social Teaching: Care for God's Creation)
6. Rubric

Some scientific facts we have learned through the observation and monitoring of these fascinating birds:

1. Eastern bluebirds are cavity nesters (birds that build nests, lay eggs, and raise young inside cavities).
2. Eggs are blue and the dimensions are 16x21 mm approximately. The female lays 5-6 eggs in a clutch.
3. They are altricial birds:
    a. They are weak and helpless for a while after hatching.
    b. When they hatch, they have no feathers, and their eyes are closed.
    c. They cannot walk or fly after hatching.
    d. Their parents feed them for several weeks and keep them warm.
4. Their nests are neat and made of fine grass.
5. Incubation time: 13-16 days.
6. Fledging Time: 18-19 days.
7. Bluebirds are perching birds, so they have special adaptations for resting on branches (their feet automatically close around the branch).

Other Considerations:

1. Entrance hole should face east or northeast to prevent sunlight from shining into the hole and overheating the box interior.
2. Nesting begins in mid-April.
3. Used nest must be removed. This allows a second brood to be raised. They renest after the first brood leaves.

The last document on this STEM integration is a **Transdisciplinary Curriculum Plan on Sheltering and Monitoring Bluebirds** that includes:

1. Overview
2. Activity goal
3. Rationale
4. Connecting to standards
5. Prior knowledge
6. Real-world context
7. Background STEM content
8. Final Project
9. Catholic Identity
10. Career Connection
11. Additional Considerations

"In the transdisciplinary approach, teachers bring together the twenty-first-century skills, knowledge, and attitudes to real-world application and problem-solving strategies" (Vazquez et al. 69). These real-world problem projects have multiple solutions where students need to master knowledge in different disciplines as well as skills. The focal point of this STEM integration was to use the engineering design process to design and build houses for the local bluebirds. Students had to consider the constraints given by the characteristics and nesting habits of the birds and all disciplines are joined in a seamless way to create habitats for these fascinating perching birds whose habitats have been destroyed due to deforestation. It is also important to have relevant speakers to the topic of discussion. After the environmental engineer's presentation, students created a job board with the following information about environmental engineers: what they do, where they might work, education, and other job requirements (Dispezio 191).

Students completed the design of six more birdhouses, and it was a parent-teacher-student effort as seventh grade parents joined us to complete the designs students engineered. Students also designed bar graphs to represent the population change over time for the Eastern bluebirds. To express their empathy for these creatures created by God, students wrote poems. Poetry was a powerful way of self-expression to close this unit. Students also reflected on their roles as stewards of God's creation.

Curriculum Alignment – STEM Integration Sheltering and Monitoring Bluebirds[4]

Task	Subject	Diocesan Objective
Raise questions about birds' biodiversity and ecosystems. Generate appropriate explanations based on those explorations.	Science	
Analyze data and scientific articles and use them as evidence to discuss how all human activities have short and long-term consequences for the ecosystems.	Science Math	
Discuss the ethical considerations that those consequences raise.	Religion	
Design and present an engineering design project: define problem, develop solution, and improve design.	Engineering	
Define the criteria and constraints of the engineering design.	Engineering	
Communicate and evaluate solutions.	Engineering	

---

[4] The STEM Integration chart template was developed by the Center for STEM Education at the University of Notre Dame ~STEM Integration Module facilitated by Dr. Gina Navoa Svarovsky. This chart was adapted to meet the needs of this STEM Integration on Eastern bluebirds.

STEM Integration Experience[5]

Activity	Description of Activity	Content and Practice Goals	Assessment Choice
Lesson 1: Contextualizing the problem	The problem bluebirds are facing in U.S. and Florida is discussed: • Students discussed the scientific article on bluebirds from the Department of Environmental Protection. • Students analyze data on bluebird population.	Science: using data to create evidence, analyzing information to make arguments from evidence. Math: data representation. Engineering: grasping the context of the real-world problem. English: analyze, summarize, and discuss scientific information.	Formative assessment
Lesson 2: Who are environmental Engineers?	• Discuss the role and job description of environmental engineers.	Engineering: discuss the work of an environmental engineer.	Formative Assessment
Lesson 3: Delimit the problem by determining the constraints and criteria of the design (it includes scientific facts and principles)	Discuss and consider: • Bird Characteristics: nesting, brooding, raising young, and fledging time. • Food pyramid for bluebirds. • Taxonomy.	Engineering: use the EDP to determine criteria for design. Science: investigate birds' characteristics, flow of energy, and taxonomy.	Summative Assessment: Test
Lesson 4: Developing Possible Solutions ~ Birdhouses for Bluebirds Optimizing Solutions	• Use the EDP to design birdhouses for bluebirds Considerations for the design: roof, ventilation, visibility, perching area, hole size, cavity depth for nesting. • Monitoring bluebirds' activity and sharing information about nesting. • Advanced Considerations: mini camera to provide real time video and audio.	Engineering: creating and testing the prototypes, optimizing solutions.	Performance Assessment Rubric
Communicate Solution	• Students share their solutions with the class. • Complete an individual reflection.	Science and Engineering: use data to evaluate solutions.	

---

[5] The STEM Integration chart template was developed by the Center for STEM Education at the University of Notre Dame ~STEM Integration Module facilitated by Dr. Gina Navoa Svarovsky. This chart was adapted to meet the needs of this STEM Integration on "Sheltering and Monitoring Bluebirds"

Next Generation Science Standards[6]

Objective MS-LS2-5: Ecosystems, Interactions, Energy, and Dynamics	Science and Engineering Practices:	Discipline Core Ideas:	Crosscutting Concepts:
Evaluate competing design solutions for maintaining biodiversity and ecosystem services	Engaging in argument from evidence Evaluate design solutions	Ecosystem Dynamics Biodiversity and Humans Developing Possible Solutions	Small changes in one part of the system might cause change in another part Science addresses questions about the natural world Influence of science, technology, and engineering on society and nature
Objective MS-ETS1: Engineering Design Define the criteria and constraints of a design problem Evaluate competing design solutions Analyze data from tests to determine similarities and differences among solutions Develop a model or prototypes	**Science and Engineering Practices:** Asking questions and defining problems Developing and using models Analyzing and Interpreting data Engaging in argument from evidence	**Discipline Core Ideas:** Defining and delimiting engineering problems Developing possible solutions Optimizing the design solutions	**Crosscutting Concepts:** Influence of Science, Engineering, and Technology on Society and the Natural World

---

[6] For more information about the Next Generation Standards, visit the website: https://www.nextgenscience.org This chart was adapted to meet the needs of this STEM Integration on Eastern bluebirds.

STEM Activity Building Birdhouses

Objective: Use the engineering design process to design birdhouses for bluebirds.

Some birds nest in tree cavities but as forests are cleared for human developments, habitats are destroyed reducing these birds' shelters. You will design and construct habitats for bluebirds that meet these birds' specific needs. Birds need shelter for protection from weather, predators, and for nesting purposes.

Materials: wood, hammer, nails, screwdriver, and wood screws.

Constraints:

1. You should design a wooden birdhouse for bluebirds.
2. The hole size should meet the bluebird requirements.
3. The birdhouse should have a well-defined cavity (depth requirement) for the birds to build their nest and eventually, lay their eggs.
4. Determine the shape of the structure of the birdhouse keeping in mind it should be attractive to the birds.
5. Make the house attractive to the birds but not to their predators (just treat the wood or stain it with materials that are designed for natural woods).

The Engineering Design Process (EDP):

Ask:
1. What is the purpose of designing birdhouses for bluebirds?
2. Why do the materials from the list will work well to design the birdhouse?

Imagine:
1. How will you design the birdhouse? Brainstorm and draw pictures of your ideas. Circle the choice that you consider the best idea.

Idea # 1	Idea # 2

Plan:
1. Draw a diagram of your model: size of the model must be included; label the different parts of your model; for each labeled part, write the material(s) you will use.

Create:
1. What did the design of the prototype help you understand about your birdhouse?
2. Which part(s) of your birdhouse worked the best?
3. Which part(s) of your birdhouse did not work well?

Improve:
1. Explain some ways you can improve your model. Think about the size of your model, the materials, and the shape.

Draw your model's improvements in the box below.

Communicate:
Prepare a presentation about your design and how you solved the problems encountered. Include at least one picture of your design.

Answer the following questions in your presentation:

1. What do you think is the best feature of your design? Why?
2. If you had more time, what design changes would you add to make your birdhouse even better?

Catholic Identity: (include this part in your presentation too).
Catholic Social Teaching: Care for God's Creation.
As Catholics, how can we respond to the destruction of habitats for animals?

"Designing a Birdhouse" Rubric[7]

Engineering Design Process	Exemplary: consistent and independent. (4)	Accomplished: consistent with little to no teacher prompting. (3)	Developing: inconsistent and/or require teacher prompting. (2)	Beginning: experiences difficulty even with teacher prompting. (1)
Ask	Carefully consider the investigation guidelines. Thoroughly discuss multiple relevant questions for further exploration.	Consider the investigation guidelines. Discuss multiple relevant questions for further exploration.	Consider the investigation guidelines. Minimal discussion of multiple relevant questions for further exploration.	Minimal consideration of the investigation guidelines. No discussion of multiple relevant questions for further exploration.
Imagine	Generate and explore advanced and unique solutions and ideas without teacher guidance. Work with the team to generate possible solutions to the team's questions and design challenge. Independently perceive and accept the team's point of view. Carefully consider the constraint of the design challenge.	Generate and explore solutions and ideas without teacher guidance. Work with the team to generate possible solutions to the team's questions and design challenge. With the teacher's help, perceive and accept the team's point of view. Consider the constraint of the design challenge.	Generate some solutions and ideas without teacher guidance. Team's work to generate possible solutions to the team's questions and design challenge with teacher guidance. Minimal consideration of constraints.	Generate some solutions and ideas without teacher guidance. Does not attempt team's work to generate possible solutions to the team's questions and design challenge. Solution does not consider the constraints of the design challenge.
Plan	Keep detailed records and sketches of design possibilities and adjustments. Use the International System of Units to sketch every detail of the design. Critically evaluate the dimensions and materials of the design. Use prior knowledge to design the prototype.	Keep records and sketches of design possibilities and adjustments. Proficient use of the International System of Units to sketch the design. Evaluate the dimensions and materials of the design. Use prior knowledge to design the prototype.	Keep records and sketches of design possibilities and adjustments. Does not use the International System of Units to sketch the design. Minimal evaluation of the dimensions and materials of the design.	Attempt to create records and sketches of design possibilities and adjustments. Does not use the International System of Units to sketch the design. No evaluation of the dimensions and materials of the design.

---

[7] This rubric was created with the basic five steps of the Engineering Design Process from the "Engineering is Elementary" program. The rubric was modified to meet this project's needs. To learn more about their STEM curricula, visit https://yes.mos.org/curricula/

Educating the Whole Child in Catholic Education

Engineering Design Process	Exemplary: consistent and independent. (4)	Accomplished: consistent with little to no teacher prompting. (3)	Developing: inconsistent and/or require teacher prompting. (2)	Beginning: experiences difficulty even with teacher prompting. (1)
Create	Persevere to create a functioning prototype. Thoughtful use of materials and resources.	Persevere to create a functioning prototype. Use materials and resources.	With teacher's help, persevere to create a functioning prototype. Basic use of materials and resources.	Lack of perseverance to create a functioning prototype even with teacher's support. The solution shows no evidence of the use of materials and resources.
Improve	Without teacher guidance, self-evaluate and critique the functioning prototype and analyze all design deficiencies. Suggest multiple ways to improve the efficiency and quality of the prototype.	With teacher guidance, self-evaluate and critique the functioning prototype and analyze all design deficiencies. Suggest ways to improve the efficiency and quality of the prototype.	With teacher guidance, self-evaluate the functioning prototype and analyze some of the design deficiencies. Some considerations to improve the efficiency and quality of the prototype are suggested.	Lack of self-evaluation of the functioning prototype and flaws. Some considerations to improve the efficiency and quality of the prototype are not attempted even after teacher's intervention.
Clarity of Communication	Present relevant visual aids to support presentation. Reflect preparation. Logical progression of ideas and supporting evidence.	Present visual aids to support presentation. Reflect preparation. Logical progression of ideas.	Visual aids somehow support presentation. Ideas are not or effectively structured.	No visual aids are included. The information presented is unfocused, poorly organized and show little effort.

Transdisciplinary Curriculum Plan on Sheltering and Monitoring Bluebirds

- **Overview**: Some birds nest in tree cavities but as we clear forests for human's development, animal habitats are destroyed. Birds need shelter for protection from weather, predators, and for nesting. We can design and construct habitats that meet birds' needs. There are many careers involved in developing shelters for animals.
- **Activity Goal**: Use the engineering design process to design birdhouses for bluebirds.
- **Rationale**: Birds need shelter for protection from weather, predators, and for nesting purposes. Deforestation has reduced the birds' habitats.
- **Connection to Standards**: unit on birds/vertebrates (science), unit on poems (literature), unit on creative writing (English), unit on graphing (math).
- **Prior Knowledge**: definition of a vertebrate, classification levels, basic needs of living things, and basic knowledge on the engineering design process.
- **Real-World Context**: Some birds nest in tree cavities but as forest cleared for human development, habitats are destroyed.
- **Background STEM Content**: How do birds' characteristics determine birdhouse design requirement? Use the engineering design process to design houses for bluebirds.
- **Final Project**: Students will use the characteristics and specific needs of bluebirds to design shelters for this type of birds. Students will use the engineering design process to design the prototype. Students will research on birds, observe fertile eggs until their successful hatching, and write about the daily development of birds. Students will be able to share information about bluebird's nesting through the nest watch website. The unit will end with students' poems on birds.
- **Catholic Identity**: Students will reflect on God's creation and how to respond, as Catholics, to the destruction of habitats for animals. As part of the analysis, students will create graph on bluebird populations to see how the destruction of habitats has impacted the population.
- **Career Connection**: Birds need shelter for protection. Birds that nest in tree cavities are losing their habitats due to deforestation. We can design houses for birds to meet their specific needs. There are many careers involved in developing shelters for animals that require knowledge and skills in technology, science, math, art, engineering, and other subjects. Some careers/jobs are wood workers, life scientists, farmers, among others. Career connection is a relevant practice as it also gives the opportunity to explain students that our work brings dignity, meaning and purpose to our lives. Any job that we perform with honesty, excellence and for the common good is our way to use our God-given talents well.
- **Some Additional Considerations**: Cedar is the best type of wood to build the birdhouse. The hole size should meet the bluebird requirements, and the birdhouse should have a well-defined cavity for the bird to build a nest and eventually lay eggs. The shape of the house and roof inclination must be carefully considered. The house must be attractive to birds but not to predators.

STEM Integration: Saving Florida Sea Turtles in their Race for Survival[8]

The Federal Water Pollution Control Act imposes limits on pollutants discharges from industrial facilities and proposes the use of pollution control technology. The pollution in the oceans is affecting marine life in general, which is affecting the lower level in the food chain (plankton.) All species of sea turtles in Florida are endangered of extinction and are being affected by marine pollution. Students conducted a controlled experiment to determine the best filter materials to clean nontoxic contaminated water and then use the Engineering Design Process (EDP) to design the filtration systems.

The goals of this integration were:

- Design a controlled experiment to explore filter materials.
- Create and interpret box and whisker plots to analyze the filtration time and water quality of the filtered water.
- Use the Engineering Design Process (EDP) to design a pollution control technology.
- Propose and communicate solutions.
- Understand the characteristics of sea turtles as reptiles, the food web of sea turtles, and the environmental problems they face in their race for survival (contextualizing the problem).

We started the integration with the Engineering is Elementary (EiE) story "Saving Salila's Turtle"[9] and a scientific article from the Sea Turtle Conservancy to understand the struggle for survival of sea turtles. Students created a poster of the Engineering Design Process (EDP), and they used paper circuitry to illuminate the icons they chose for steps of the EDP. While students designed the controlled experiment and created the box and whisker plots to analyze the materials, they also discussed the federal efforts to save the Florida sea turtles. They wrote letters to a Florida senator to propose solutions. The cost of the materials and the information from box and whisker plots gave students the starting point to work on their designs. They used the EDP to design the filtration systems and communicate the results to the rest of the groups and to their parents through our "Engineering Showcase Day."

---

[8] Saving Florida Sea Turtles in their Race for Survival was inspired on the STEM Integration on filtration systems to address the Flint Water Case ~ Center for STEM Integration ~ Module 2, facilitated by Gina Svarovsky ~ University of Notre Dame
[9] To learn more about EiE modules, visit https://yes.mos.org/curricula/

My assessment plan was as follows:

Formative assessments: Students had different check points during the integration. They were:

- Exploring filter materials (controlled experiment): hypothesis and research explanation.
- Results: sharing results on Google classroom file.
- Interpreting box and whiskers plots: using the information from the controlled experiment as evidence to design the filtration system.
- Engineering Design Process: presenting four possible solutions and the plan for the final solution (drawings, labels, dimensions, cost, justifying their solution).
- Communicating their final design to the class.

Summative Assessment: engineering reflection to assess box and whisker plot interpretation (building explanation by using evidence), designing controlled experiments, applying the EDP, and proposing solutions (decision making skills). A grading scale and a holistic rubric were used to assess students' performance. Adding a rubric and a scale to the engineering reflection made the assessment very effective. Some of the characteristics that made the rubric effective were the use of categories, level descriptors, weighting, scales, and the use of factors to multiply the categories that were essential to the unit goal (the engineering design process and proposing solutions). More detailed information on holistic assessments is found in Chapter 5.

To close the STEM integration, we attended the SeaWorld Educational Program "Saving Our Seas." After the field trip, students created brochures to educate people on how to be part of the solution for marine pollution. They also wrote books about how to save our seas by positioning the readers as helpful humans. We have used the Story Jumper website to create and publish their illustrated books. It was valuable to host an environmental engineer as a speaker to make the career connection. Students were very engaged during the 45 minutes presentation and asked meaningful questions to understand the job of an environmental engineer.

Overall, the experience was successful. Students were excited and engaged. The "Engineering Showcase Day" was another layer for communicating solutions and a way to create STEM awareness in our community. In addition to the scientific article from the Sea Turtle Conservancy[10], real data from Sea World[11] was used to contextualize the problem.

---

[10] To access the Sea Turtle Conservancy webpage, visit https://conserveturtles.org

[11] Visit https://seaworld.com/educational-resources/sea-turtles/ for useful information.

One of the successes was the "Engineering Showcase Day" where students were able to communicate their solutions to their parents. Another important success was the opportunity to connect the scientific method and the engineering design process. Students had to design a controlled experiment to explore the filter materials (filtration time and quality of water) and then they used the EDP to engineer a filtration system.

For this integration, I have included the letter we sent to parents before we started the integration as well as the unit plan. I cannot stress enough the value of communicating with parents when project-based learning is taking place. I have lost count on the many opportunities our parents have shared that those letters allowed them to initiate a dialogue with their children. In addition to that, it is the most effective way to find speakers for most of the STEM integrations. The speakers bring the career connection which is an important goal of this type of approaches. The unit plan contains:

- Connection to Catholic Identity
- Essential Questions
- Unit Summary
- Standards Addressed
- Learning Outcomes
- Cross-Curricular Collaboration
- Unit Implementation
- Accommodations

This STEM Integration was done with the cooperation of our wonderful middle school cluster teachers.

October 6th, 2017

Dear Parents,

Next week, your child will be part of the STEM integration "Saving Florida Sea Turtles in their Race for Survival." To contextualize the problem sea turtles are facing in U.S. and specifically in Florida, we will approach this topic from several angles:

1. Discussing the environmental obstacles these beautiful reptiles face in their race for survival.
2. Analyzing Sea World data about threatened and endangered sea animals.
3. Researching the Endangered Species Act.
4. Introducing students to environmental engineering through the story "Salila's Turtle."
5. Identifying in the story the use of the Engineering Design Process and creating an illuminated diagram of the process.
6. Understanding turtles' taxonomy and creating identification cards of Florida sea turtles with the conceptual model about the levels of classification of living things.

The engineering challenge comes from the Engineering is Elementary (EiE) module "Designing Filtration Systems," in which students will be combining the engineering design process and the scientific method to design and improve a pollution control technology for industrial wastewater treatment:

1. Design a controlled experiment to determine the best filter materials (we will be using non-toxic contaminated water).
2. Using box and whiskers plot to represent and interpret the data.
3. Using the steps of the engineering design process to design the filtration system.

We are planning an "Engineering Showcase Day" where parents will be able to support our local young environmental engineers as they share their engineered filtration systems. The date will be announced later this month. After we come back from our educational field trip to Sea World, students will be designing a conservation campaign brochure and a children's book to create awareness of the problem sea turtles are facing and what we can do to help them on this quest, under the frame of the Catholic Social Teaching "Care for God's Creation" as a requirement of our faith. The enduring understanding of this project is that humans have an impact in the environment and that our major way to save our planet is by practicing conservation. Our most distinctive mark at Catholic schools is our Catholic Identity and we cannot emphasize enough our role at the top of God's creation. Our speaker for this integration will be an environmental engineer who will explain the job description of a professional in this field and how they address problems of soil, air, and water contamination.

Later this year, we will be designing prosthetic tails for fish as part of our unit on biotechnology. This second module "A Deep Dive into Prosthetic Tails," is also part of the EiE program that we purchased thanks to a generous contribution from one of our families. Our deep appreciation for this blessing, which gives us the opportunity to design meaningful STEM integrations that will contribute to the formation of the minds and conscience of our youth, cannot be understated.

7th/ 8th Grade Cluster

Unit Plan ~ STEM Integration: Saving Florida Sea Turtles

Name: Yanny Salom	Grade: 7th
Subject Area: Life Science	Quarter: 2nd
Anticipated Dates of Instruction: 10/13 – 11/1	

Title	STEM Integration: Saving Florida Sea Turtles

Connection to Catholic Identity

Catholic Social Teaching: Care for God's Creation as a requirement of our faith.

Essential Questions

1. How do turtle's nesting characteristics and the environmental problem they face can be used as evidence to design a conservation campaign in the race of survival of Florida sea turtles?
2. What are the traits that allow reptiles to live on land?
3. What are the levels of classification and scientific names of the Florida sea turtles?
4. What is the energy flow of the marine ecosystem of sea turtles?
5. Identify characteristics of reptiles as vertebrates.

Unit Summary

Human actions have an impact in our environment and the living organisms. There are patterns of interactions among organisms across multiple ecosystems that is affecting the flow of energy among living and nonliving organism. One major way to save our planet is by practicing conservation. We can save sea turtles by promoting awareness and by adopting them through the WWF. There are many careers involved in conservation: engineers, ecologists, life scientists, website designers, etc.

Standards Addressed

Learning Outcomes

Transfer: How will student be able to independently use this learning in the future?

Understanding that one major way to save our planet is by practicing conservation. The Endangered Species Act effort to protect our world.

Meaning: What important ideas and inferences do you want students to grasp and understand?

Awareness of the race for survival of Florida sea turtles and the creation of a campaign to be part of the solution.

Designing a controlled experiment to determine the best filter materials to design filtration systems.

Acquisition: What facts and concepts should students know and be able to recall?

Taxonomy of living things, characteristics of reptiles, endangered species conservation act, steps of the engineering design process and the scientific method, ways to help and be part of the solution.

Acquisition: What discrete skills and processes should students be able to use?

Designing campaigns and writing books to create awareness of the environmental problems the sea turtles are facing.

Combining the scientific method and the engineering design process to design solutions.

Cross-Curricular Collaboration

Math: Using Box and Whiskers Plots to represent data, calculating percentages and bar graphs
Art: Creating icons for the Engineering Design Process (EDP)
Social Studies: Endangered Species Act
Religion: Catholic Social Teaching
Literature: Fiction Story "Salila's Turtle" from the EiE program
Media: Designing brochure and story book
STEM: Using paper circuitry to illuminate the steps of the EDP, designing and improving a pollution control technology (filtration systems)
Field Trip to Sea World
Career Connection: Environmental engineer speaker

Instructional Strategies	Grouping Options	Resources/Assignments
Directed Reading on Classification of Living Things	Independently	Textbook: module *Living Things*
Class discussion on directed reading assignment	Whole Group	Directed Reading Assignment
Cornell notes on classification of living things	Independently	Cornell Notes template
Section Review on Classification of Living Things Feedback on section review	Whole class	Section Review Answers and book

Educating the Whole Child in Catholic Education

Instructional Strategies	Grouping Options	Resources/Assignments
Quiz on Classification of Living Things	Independently	Quiz
Students receive their quiz back and we go over the answers. Allow students to ask questions and clarify ideas. Test and quizzes must be placed in science binder and become part of study material for final test.	Whole group	Graded quiz
Assignment: Identification cards of Florida sea turtles	Independently	Computer, printer, book, computer paper
Test on classification of living things	Independently	Graded test
Students receive their test back and we go over the answers. Allow students to ask questions and clarify ideas. Test and quizzes must be placed in science binder and become part of study material for final test.	Whole class	
Directed Reading on Reptiles	Independently	Old Science textbook p.426-430
Class discussion on directed reading assignment	Whole group	Directed Reading Assignment
Cornell Notes on Reptiles	Independently	Cornell Notes template
Section Review on Reptiles

Feedback on section review | Whole Class | Section review assignment and book |
| Quiz on reptiles | Independently | Graded quiz |
| Students receive their quiz back and we go over the answers. Allow students to ask questions and clarify ideas. Test and quizzes must be placed in science binder and become part of study material for final test. | Whole group | |

Instructional Strategies	Grouping Options	Resources/Assignments
Scientific article on the challenges the sea turtles face in their race for survival.	Small groups	Article, reading comprehension questions
Test on reptiles and scientific article.	Independently	Graded test
Students receive their test back and we go over the answers. Allow students to ask questions and clarify ideas. Test and quizzes must be placed in science binder and become part of study material for final test.	Small groups	
Students read the fiction story "Saving Salila's Turtle" and are introduced to the Engineering Design Process (EDP) ~ Literature Class	Small groups	Story from EiE
Students read about the EDP	Independently	eTextbook
Students create icons of the EDP and illuminate them with paper circuitry~ Art Class	Small groups	Paper circuitry kit
Presentation on food webs of sea turtles	Whole class	Power point presentation
Designing food pyramids based on food web diagrams	Independently	Worksheet with directions, rubric, and food web
Students receive feedback on the assignment	Independently	Rubric with feedback
Designing a controlled experiment to explore filter materials. Students will create a folder with a cover (name of their company must be included).	Small groups	Worksheet Power Point presentation Folder, computer papers, shared worksheets
Running the experiment to collaborate on the creation of the data. Create box and whisker plots to represent the data ~ math class.	Small groups	Worksheet from Google Classroom Box and Whisker Worksheets EiE Kit/Module

Instructional Strategies	Grouping Options	Resources/Assignments
Each group/company will analyze the box and whisker plots to create a filtration system (the challenge). Constraints are discussed and possible trade off.	Small groups	EiE module EDP worksheets
Each group will submit a final report with the justification and description of their solutions.	Small Groups	Report
Groups receive their report back and we go over the answers. Allow students to ask questions and clarify ideas.	Small Group	Report with feedback
Independent Engineering Design Process Reflection	Independently	Engineering Reflection Worksheet
Students receive their engineering design process reflection back and we go over the answers. Allow students to ask questions and clarify ideas. Reflection must be placed in their science binder (tab: STEM Integrations) and become part of study material for final test.	Independently	Graded Engineering Reflection
Reflection on our role as stewards of creation ~ Religion Class Writing books to create awareness about endangered species with Catholic component ~ Media and Religion Classes Research the Endangered Species Act ~ Social Studies		

ADHD/Dyslexic Processing Autistic/Delayed	Check Cornell Notes to be sure notes are accurate. Provide extra help on creating Cornell Notes. If needed, allow them to come early in the morning to check understanding. Allow extra time for directed reading assignments. Use audio to listen to the lessons.
Challenged Students	Call on students frequently to check for understanding. Check Cornell Notes for accuracy. Encourage the students to find evidence in their textbooks when providing answers.
Accelerated Students	Look for more detailed answers in questions. Encourage the students to find evidence in their textbooks and in other sources when providing answers. Allow students to help you organize the laboratory activity.

Chapter 4

We Are Better Together

> "Two are better than one, because they have a good reward for their toil.
> For if they fall, one will lift up the other, woe to one who is alone
> and falls and does not have another to help."
>
> Ecclesiastes 4, 9-10

Integrating subjects allows students to make meaningful connections and understand the relevance of the content they are learning. STEM integration goes beyond the traditional disciplines of science, technology, engineering, and mathematics and includes other subject areas to make learning more impactful and relevant. STEM is a way to organize the curriculum in order to deliver instruction in a meaningful way. Middle school cluster teachers had the desire and disposition to integrate disciplines through project-based approach. We aligned the projects with the curriculum we were teaching and made the necessary adjustments to deliver instruction. We saw the benefits of collaboration through the contagious enthusiasm of our students. Their eagerness to execute the projects and their happiness upon witnessing the connection among disciplines gave them a reason to understand why they were learning the content in each subject area. Applying the concepts or skills learned in a real-life project create a connection between the knowledge learned and how it can be applied in daily life. It also elevates students' self-motivation and confidence. Educating the whole child requires students to be engaged, motivated, supported, and challenged in a safe environment that fosters the development of communication, problem-solving, critical thinking, and social skills.

Our project-based approach allowed us to contextualize the learning material by connecting to real-world situations. Students were applying concepts to show understanding rather than learning by heart without seeing a real application or having exposure to a real-life situation. One additional benefit of this approach was to build students' confidence. Students struggle to see failure as part of the learning process; finding opportunities to apply knowledge and using skills are effective strategies. This type of approach was very helpful for students with learning differences because more time was offered to provide solutions. It was also helpful for advanced students who had the opportunity to attack problems with multiple solutions where risk/benefit analysis were our daily practice. This was the perfect formula for challenging the accelerated and accelerating the challenged. After all, the teaching profession is a daily balancing act, with the goal of reaching students exactly where they are.

The following projects illustrate the value of working together: designing roller coasters, STEM books, constellations, and the free market project.

Roller Coasters

One of the projects included in this section is "Designing Roller Coasters." This type of project is a great opportunity to maintain engagement and to continue building rapport with students. The project not only facilitates the integration of various disciplines but also allows for the opportunity to practice collaboration and respect among students. In Catholic education, this holistic and practical approach also considers the spiritual development of students. This project allows students to work in conjunction with their religion teacher to write a reflection on how the journey of faith should not be analogous to the ups and downs of a roller coaster ride.

This project follows Stacey Green's five principles of integration (Vazquez et al. 18) as follows:

Focus on Integration: This project was done after teaching motion, forces, and energy because this way students were able to use learned concepts to create their prototype. The design and construction of the model required that students asked, imagined, and planned solutions that were then created, tested, and improved. In addition to that, students recorded data to calculate the speed of the rider and the gravitational potential energy in different points of the roller coaster model. The use of complex technology was required to create an iMovie about the history of roller coasters or America's first amusement park. Students also designed virtual roller coasters, which taught them about the importance of friction, energy transfer, and the law of conservation of energy before designing the physical model of the roller coaster, for which they needed to use simple technology. We integrated our faith by reflecting on how to avoid spiritual roller coasters when our relationship with God seems to go up and down.

Establish Relevance: Students were positioned as roller coasters designers who were hired to design a roller coaster for an amusement park. Students also researched on magnetic levitation technology (maglev technology) to find ways to make roller coaster rides even more exciting.

Emphasize Twenty-First Century Skills: Students practiced communication and collaboration skills by debating, discussing, supporting claims, and arguing from evidence. It was inspirational to listen to students' discussions on how to improve methods and designs showing great respect to each other and based on scientific facts and concepts. This type of approach gives the opportunity to practice critical thinking and social skills that will be so needed in students' personal and professional endeavors.

Challenge Your Students: Students applied what they learned in science and math to plan, design, construct, test, and improve a roller coaster model. The design is based on data about variables like height, number of hills, number of loops, among others. The beauty of this approach is the opportunity to evaluate the design arguing from evidence.

Mix It Up: The project gave students the opportunity to demonstrate understanding of content and skills in different disciplines, presenting ideas, and drawing conclusions. The "STEM Open House" events were a great opportunity for students to demonstrate their prototypes and explain how they arrived at their designs by applying concepts and skills.

The following documents have been included to illustrate this STEM integration:

- Catholic Identity
- iMovie: History of Roller Coasters or America's First Amusement Park
- Rubric ~ iMovie
- Six Traits of Writing on new technologies to make roller coaster rides more sensational and faster
- Thrilling Ride: Designing Roller Coasters ~ STEM Lab Activity
- Rubric ~ STEM Lab Activity

The "Six Traits of Writing" rubric was a school wide implementation under the leadership of the English and literature teachers. This rubric initiative provided clear expectations about the writing skills, and it also facilitated peer-reviews. In addition, teachers consistently assessed and evaluated the quality of students' work. Verbal and written communication skills are important personal and professional skills that we must foster and develop in our students. This enables students to build assertive relationships, positive interactions, clear understanding of professional and personal matters, and the ability to communicate in work and social situations in an effective way. Language is one of the most important human attributes. I moved to the United States of America twenty-five years ago with the ability to read and comprehend technical English, but verbal communication was a path I had to develop here. I spent all these years asking students to love their English language. Working on improving their verbal and written communication skills is a way to enhance and honor their language, which is also part of their cultural identity. The Six Traits of Writing Rubric[12] was based on the following categories:

- Ideas and Content (main theme and supporting details)
- Organization (introduction, structure, and conclusion)
- Voice (personality and sense of audience)
- Word Choice (precision, imagery, and effectiveness)
- Sentence Fluency (flow, variety, and rhythm)
- Conventions (age-appropriate spelling, caps, punctuation, and grammar)

---

[12] The Six Traits of Writing Rubric was a school wide assessment that was designed under the leadership of the English and Literature Department.

Designing Roller Coasters

Catholic Identity:

Just like roller coasters, your personal growth in faith may have highs and lows. You may feel happy when you are up as you do in the highest point of the roller coaster. You may feel sad and disheartened when you start going down in your faith. Stay always closer to God, in your peaks and valleys.

Does your relationship with God seems to go up and down? What do you need to do to avoid the spiritual roller coaster? Remember, a roller coaster can be fun in an amusement park but not in your spiritual life. God promises to be with us in the ups and downs of life. "And remember, I am with you always, to the end of the age." **Matthew 28:20.**

Performance Based Assessment (Small Group Activity)

iMovie: History of Roller Coasters or America's first Amusement Park.

Roller coasters were not born in amusement parks. The first roller coaster in United States was not created for amusement, but for hauling coal. It was known as the Mauch Chunk Switchback Railway in eastern Pennsylvania. Visit the "Coney Island" page of *PBS's American Experience*[13] to learn about the history of roller coasters or America's First Amusement Park. Then create a documentary using the iMovie app to teach others about the chosen topic.

Authentic Setting: Media Center and Science Laboratory

Role: Movie Producer

Audience: Students in your class

Performance - Based Assessment (Small Group Activity)

Six Traits of Writing on new technologies to make roller coaster rides more sensational and faster.

Gravity and momentum take over the roller coaster rides to make the ride a stimulating experience. However, new technology is being used to gain acceleration. Research maglev technology and explain what it is about. How can roller coasters use this technology to make the rides a more breathtaking experience?

Authentic Setting: Media Center and Science Lab

Role: Writer

Audience: Physical Science Teacher

---

[13] To learn about the history of roller coaster in America, visit https://pbs.org/wgbh/americanexperience/features/coney-century-screams/

iMovie Rubric

History of Roller Coasters or America's First Amusement Park

Category	3	2	1
Content: • **History of Roller coasters or** • **America's first Amusement Park**	The movie delivers content about the history of the roller coasters or America's first amusement park. The movie exceeds the expectations to deliver the content.	The movie delivers content about the history of the roller coasters or America's first amusement park. However, the content has been oversimplified.	The movie does not deliver the content at all or the content does not match the choices offered by the teacher.
Visual: • **Photos/Images** • **Visual Effects**	Movie contains good quality visual images with appropriate visual effects.	Movie contains poor quality images with visual effects that distract viewer from the content.	Movie does not contain good images and in small number so the images appear for too long.
Audio: • **Clear** • **Appropriate to the content**	Audio is clear and appropriate to the content.	Audio does not meet one of the requirements listed.	Audio does not meet the requirements listed.
Use of Class Time: **Student** • **Stays engaged during activity** • **Makes meaningful contributions to the project.** • **Does not disturb other groups**	Meet all the listed requirements under the criteria "use of class time."	Meet at least two of the listed requirements. Student may receive a classroom offense for not meeting one of the requirements under the criteria "use of class time."	Student has received several warnings about meeting the requirements under "use of class time". A classroom offense was issued.
Total Points / Grade			

Performance – Based Assessment

A Thrilling Ride: Designing Roller Coasters (STEM Lab / Small Groups)

Roller coaster designers decide the number of turns, loops, and twists a roller coaster may have. They also decide the height of the tallest hill, number of hills, and the length of the ride. If you want to be a roller coaster designer, you should study to be an engineer, architect, mathematician, or physicist.

Use duct tape and pipe insulation to design your own roller coaster. You will apply the concept of energy transformation, friction, law of conservation of energy, and types of energy to design your roller coaster. Your constraints will be as follows:

The maximum length of the ride will be 1.8 meters

The rider will be a marble, and it will begin its journey at the top of the highest hill.

Minimum number of hills: 2

Hill Maximum Height: 1 meter

Minimum number of loops: 1

You can make turns and twists but make sure you do not reduce the area of the pipe insulation to the point the rider (marble) cannot complete its route.

Authentic Setting: Science Laboratory

Role: Roller Coaster Designer

Audience: Students and teacher

STEM Activity

A Thrilling Ride: Designing Roller Coasters
Use your e-book or hard copy book to define the following terms. Complete the chart by applying what you know about each term and create a strategy to help you design the roller coaster. Two cells were filled out as an example.

Educating the Whole Child in Catholic Education

Terms Chart

Term	Definition	Application/Strategy
Energy		
Kinetic Energy		
Potential Energy	Energy due to position, condition, or chemical composition.	The higher the hill, the more potential energy. Part of potential energy will become kinetic energy.
Friction		
Gravitational Potential Energy		
Gravity		
Energy Transformation		
Mechanical Engineer		
Law of Conservation of Energy		

Keep in Mind the Constraints!

- Your rider must complete the entire track without stopping.
- Do not forget the constraints for your design.
- You cannot add additional energy to help the rider complete the route: It must travel the length of the track and climb over the hills. Once you have created the basic track, you can include turns and loops.
- When modifying your track, change one variable at a time (height of hills, loops, distance between hills, etc.). This practice will give you a clear picture of what may be affecting the performance in your roller coaster.
- You must draw the design plans (sketch of your roller coaster track). Include height of the hills, distance between hills, turns, and loops. Think metrically!
- Measure the mass of the marble in grams. Then, make a conversion from grams to kilograms (1Kg = 1000 grams).

Data:

1. Mass of marble in grams: _____

2. Mass of marble in kilograms: _____

3. Height of the first hill (m): _____

4. Height of the second hill (m): _____

5. Height of additional hills, if you have more than two (m): _____.

6. Potential energy of the hills: calculate it by multiplying mass (kg) by gravity (10 m/s/s) by height of the hill (m)

6.1 Hill#1(joules): _____

6.2 Hill #2 (joules): _____

6.3 Hill # 3 (joules): _____ (if you have more than two)

7. If we ignore the friction between the marble and the track, the velocity of the rider can be calculated as $V = (2 \cdot g \cdot h)^{(1/2)}$ where 2 is a constant, g is 10 m/s/s, and h is the height of the hill.

7.1 Velocity at hill #1 (m/s): _____

7.2 Velocity at hill # 2(m/s): _____

7.3 Velocity at hill # 3(m/s): _____ (if you have more than two)

8. Number of loops: _____

9. Distance between hills (cm): _____

10. Additional Information (if any):

Describing your Roller Coaster: Apply the key terms to describe your roller coaster (see chart for terms). Use at least seven sentences to describe your roller coaster. Describe all modifications that you made to your roller coaster. Make sure you share your understanding of energy transformation, friction, and the law of conservation of energy.

STEM Lab Report Rubric

Category	4	3	2	1
Conducting an investigation	You conduct the experiment safely. Make accurate observations and measurements. You identify and control variables. Present experimental results clearly.	You conduct the experiment safely. Your observations and measurements are usually accurate. You identify variables and attempt to control variables. Present experimental results clearly.	You conduct the experiment safely. You do not make accurate observations and measurements or do not control identified variables. Present experimental results clearly.	You do not conduct the experiment safely. Your observations and measurements are often incorrect because you do not control variables or the results are not clear.
Building an explanation	You communicate results and support explanations and conclusions reflecting the facts and concepts of science. You use data and graphs from the experiment to support conclusions.	You communicate results and attempt to use evidence to support your ideas and conclusions.	You attempt to communicate results. Your conclusions do not reflect the science concepts introduced and the use of data from the experiment.	You attempt to communicate results. Your conclusions are unrelated to the science facts introduced and the results obtained in the experiment.
Journal	The journal is completed correctly and contains all the parts required. It also contains additional, unexpected, or outstanding features.	The journal is partly correct. It has no big mistakes. All the parts are included.	The journal contains big mistakes. Not all the parts are included but give information that is related.	The student does not do the journal or gives answers that have nothing to do with what was asked.

Working in small group means that you: stay on task, make meaningful contribution to your group, and do not disturb others.

Educating the Whole Child in Catholic Education

STEM Books

To introduce children to the exciting world of curiosity, discovery, and real-world context; graphic novels based on true stories are a meaningful comprehensive approach. Graphic novels contain captivating visuals that help students engage easily in the stories as they also increase understanding. When they address real-world problems, it creates awareness beyond the classroom walls and bridge the gap between knowledge and providing solutions. We created a team of teachers and administrators to select the books, and we started with the following books:

For eighth grade ~ *Saving Sorya: Chang and the Sun Bear*, written by Trang Nguyen and illustrated by Jeet Zdung.

For 7th grade ~ *Fred & Marjorie: A Doctor, A Dog, and the Discovery of the Insulin*, written by Deborah Kerbel and illustrated by Angela Poon.

For 6th grade ~ *Dr. Fauci: How a Boy from Brooklyn Became America's Doctor*, written by Kate Messner and illustrated by Alexandra Bye.

Criteria for STEM Books[14]

Criteria	Saving Sorya: Chang and the Sun Bear	Fred & Marjorie: A Doctor, A Dog, and the Discovery of the Insulin	Dr. Fauci: How a Boy from Brooklyn Became America's Doctor
Engage students in thinking about and using problem solving	A young wildlife conservationist and her struggles to protect and save a sun bear that was rescued from illegal bear trade. Our role as stewards of all God has created.	Frederick Banting designed an experiment to determine if the mysterious secretion from the pancreas could be isolated to treat diabetes. The test subject were street dogs, so students discussed and reflected on the use of animals for research in the light of our Catholic faith (Catechism of the Catholic Church, 2415 & 2416).	Personal and scientific skills that led a boy from Brooklyn to become a doctor. Using scientific skills to contain the outbreak of infectious diseases and advise people on health issues. Students also reflected on the Precepts of the Church to respond to the importance of attending Mass, as Catholics. It was gratifying to hear a group recalled priests' homilies on the importance of attending Mass on Sundays and days of obligation.

---

[14] To learn more about the criteria for STEM books, visit https://www.nsta.org/blog/what-makes-good-stem-trade-book and read this National Science Teaching Association article written by Carrie Launius and Christine Anne Royce.

Integrate two or more of the STEM fields	Students created an enlarged novel page incorporating different elements to show understanding of the novel, characters, conservation, use of GPS technology and religious component (Catholic Social Teaching ~ Care for God's Creation).	Students created an enlarged novel page incorporating different elements to show understanding of the novel, characters, controlled experiment that was designed in the discovery of the insulin, and the position of the Catholic Church regarding using animals in scientific research.	Students created an enlarged novel page incorporating different elements to show understanding of the story, characters, relevant habits of mind for scientists, how our body's defense system works, and the religious component (Precepts of the Church).
Present relevant and interconnected topics	The story allowed students to learn about sun bears and their role in the forest. Also, the efforts made by conservationists to protect this species and wildlife in general. Our position as "helpful humans" was also incorporated as a relevant topic as well as careers in conservation.	Before the discovery of insulin, the diagnosis of juvenile diabetes led to death. The discovery of insulin changed everything as it became a life-saving treatment for diabetes. Students included facts about diabetes to create awareness and empathy about this condition in our school community.	This story paves the way for future discussions students will encounter in the seventh-grade science program about how a healthy immune system, good nutrition, and physical activity are key practices in the maintenance of the human body.
Make connections to the real world	GPS technology to track and collect data regarding wildlife as well as wildlife conservationist as a career.	Our speaker was an endocrinologist who was able to bring the medical perspective: diabetes as a medical condition, empathy, awareness, and the dilemma of using animals for research.	The story includes a section about how vaccines work and the different phases they must go to ensure it is safe and good at preventing illness. The immune system can learn how to protect itself from a disease by battling it or through vaccines that teach the immune system how to fight specific germs.

Help students connect both content and practices or habits of mind used by STEM fields	Teamwork, cooperation, evidence-based discussion, and creativity. Choosing quotes, scenes, narration, feelings, and thoughts from the characters. Artistically rendering the book/scenes in comic book form.	Teamwork, cooperation, evidence – based discussion, and creativity. Choosing quotes, scenes, narration, feelings, and thoughts from the characters. Artistically rendering the book/scenes in comic book form.	Teamwork, cooperation, evidence – based discussion, and creativity. Choosing quotes, scenes, narration, feelings, and thoughts from the characters. Artistically rendering the book/scenes in comic book form.

The following documents are included as a reference of how we addressed the STEM Books Graphic Novels:

- Letter to Parents (letter was sent home to parents and guardians before starting the STEM integration).
- Description of Project for 7th and 8th grade graphic novels. Sixth grade biography was implemented in the fourth quarter to respect the alignment and timing for 6th grade curriculum.
- Rubric.

September 27th, 2022

Dear Parents,

This month, we will start an interdisciplinary extension with graphic novels. Seventh grade students will be reading *Fred & Marjorie* by Deborah Kerbel while eighth grade students will be reading *Saving Soria* by Trang Nguyen. This is an effective way to develop literacy skills while learning STEM content as we show the seamless and meaningful connection among disciplines in STEM education. Special thanks to Mrs. Chakhtoura and Ms. Mascaro for integrating STEM into the ELA curriculum with interesting stories, amazing illustrations, and captivating characters. We have chosen the books from the National Science Teaching Association list "Best STEM Books K-12".

The seventh-grade graphic novel tells the true story of the discovery of insulin with a focus on how to design a controlled experiment and the use of animals for medical and scientific experimentation. We will discuss this important aspect of our faith with valuable references from the *Catechism of the Catholic Church* that clearly states that "medical and scientific experimentation on animals is a morally acceptable practice if it remains within reasonable limits and contributes to caring for or saving human life" (Catechism of the Catholic Church, 2417 & 2418).

The eighth-grade graphic novel is also based on a true story where a young conservationist is challenged to return a sun bear to its natural habitat. We will have the opportunity to review important scientific concepts on classification of living things and the use of GPS technology for wildlife conservation. We again connect with our faith by discussing one of the themes of the Catholic Social Teaching "Care for God's Creation." This is a meaningful opportunity to position students as helpful humans and stewards of all God has created.

To continue bringing career awareness to our STEM program, we need speakers. November is National Diabetes Month, and you may want to bring awareness about this condition to our seventh-grade classes. If you have any topic on conservation that you would like to share with eighth grade students, you are most welcome too. Please contact Mrs. Salom to coordinate a 30-minute talk with the classes during STEM time.

This month, we will celebrate the feast of St. Francis of Assisi, and this is a great opportunity to recall his gentle ways with animals. He is the patron saint of animals and ecology, but beyond animals, this beautiful saint of Assisi is well known for his care for the poor and the sick. After his radical conversion, he could no longer live in minimum obligations to be a follower of Christ, and he adopted the Gospel as a way of life. His values of minority, peace, brotherhood, goodness, poverty, and prayer frame very well his desire to serve rather than to be served. Let's continue our faith journey emulating those Franciscan values and living in the awareness that our words and deeds may be the only Gospel some of our brothers in Christ will ever read.

7th and 8th Grade Teachers
Literature, English, Science, and STEM

Description of Project [15]

Graphic Novel "Page Poster"

Task: You will create an enlarged graphic novel page incorporating different graphic novel elements to show a deeper understanding of the novel, characters, scientific connection, and religious components.

Requirements/Grade:

The number of graphic novel elements you incorporate and design, if done well, can affect your grade:

- 7 or more graphic elements for an A range grade
- 5-6 graphic elements for a B range grade
- 3-4 graphic elements for a C range grade
- **Panels do not count as an element***

Conditions:
- Draw panels (at least 4 different panels)
- Use creative fonts and stylings
- Color/Shade all panels and contents
- Include text which makes it clear as to what is happening in the scene(s)
- Must include at least 2 illustrations/drawings/images

Required Components:

1. Main character's job
2. Main character's passion (does it align *exactly* with their job or is it its own entity?)
3. Identify THREE thoughtful, specific character traits to describe your main character
   a. Must include three corresponding examples
   b. Examples can be in the form of illustrations, quotes, word art, etc.
4. What barriers did the character face to reach their success?
5. Outline steps taken in their research/process/experiment/experience to reach success
   a. Think: people, environments, attitudes, etc.
6. Science/STEM
   a. 7th:
      i. Draw the different elements of the controlled experiment that was designed in the discovery of the insulin: independent variable, dependent variable, constants/controls, experimental group(s), and control group(s). Include the purpose of the research.
      ii. Diabetes Awareness: come up with at least two facts about diabetes that can help our school community understand more about this disease. Choose one way you can create awareness. You can choose creating support, detecting prediabetes and diabetes, keep learning about diabetes, how to be empathetic

---

[15] Mrs. Meghan Chakhtoura masterfully put together the project description after we discussed the contribution of each discipline.

with people with diabetes, and activities to do during diabetes awareness month. For the chosen one, what are two ways you can help to create awareness?

    b. 8th:
        i. Choose a wild animal and write the classification of the chosen animal (taxonomy levels: domain, kingdom, phylum, class, order, family, genus, and species.) After classifying it, write a short paragraph explaining the threats to the animal you chose.

You can find some options of wild animals in the World Wild Foundation page at:

https://gifts.worldwildlife.org/gift-center/gifts/Species-Adoptions

        ii. Explain how GPS technology can help track and collect data regarding wildlife.

7. Religion: Catholic Identity Component
    a. 7th: Reflect on the use of animals for research in the light of our Catholic faith. Refer to the *Catechism of the Catholic Church* (CCC 2415 & 2416). Some guiding questions:
        i. What is the position of the Catholic Church regarding using animals in scientific research?
        ii. Based on the story, was this morally correct? Explain your answer.
    b. 8th Religion: As a practicing Catholic, explain how you would respond as good stewards of God's creation considering the threats that human interactions and practices pose to these species.

Materials Needed:
- One large 'paper poster' supplied to you
- Markers, crayons, colored pencils

Design Choices:
- Separate panels each with different elements/elements.
- Groups of panels showing different elements/scenes.
- Blend clip art, magazine cuttings neatly with your own drawings.

Skills:
- Recalling and choosing the most important scenes from your novel.
- Choosing important quotes from characters in those scenes (speech bubbles).
- Choosing important thoughts/feelings from characters (thought bubbles).
- Choosing important narration in each scene (captions).
- Artistically rendering the book/scenes in comic book form (effort is the key!).

STEM BOOKS RUBRIC[16]

CATEGORIES	10	7	5	3
Basic elements	7 graphic novel elements; There is a clear and effective use of various graphic novel elements explaining the relation to the book.	5-6 graphic novel elements; Various graphic novel elements are correctly used explaining the relation to the book.	3-4 graphic novel elements; Graphic novel elements are used explaining the relation to the book, sometimes incorrectly or out of place.	Less than 2 graphic novel elements; Graphic novel elements explaining the relation to the book, sometimes clearly incorrectly or out of place and thus confusing to the reader.
Panels/layout	4 panels included; All four panels are complete in a creative, organized format.	3 panels included; All 3 panels are complete in a creative, organized format.	2 panels included; Unorganized, hard to read and understand	
Illustration	Strip contains two completed (colored) drawings in a creative, organized format.	Strip contains one completed (colored) drawing.		
Creativity	It is obvious that the student has put a great deal of thought into the lay-out and the creating of the detailed illustrations of the Graphic Novel/ Comic; his or her ideas are original, complex and 'out- of the-box'. The graphic novel/ comic is neatly completed.	The student has thought about the lay-out, understands the book and has used his/her imagination. The student has composed quite an original Graphic Novel/ Comic; with creative illustrations and an eye for details/ descriptions, but some elements may not be excellent. The graphic novel/ comic is completed but does not always look attractive and neat.	The student has made an attempt at using his/her imagination in creating the Graphic Novel/ Comic, although it is rather unoriginal; it did not quite work out. The graphic novel/ comic is completed but does not look attractive and neat.	The student has composed an unoriginal piece of work; there is little evidence of imagination/ creativity and the lay-out does not fit the lay-out of a Graphic Novel/ Comic. Little or no effort is shown. The graphic novel/ comic is not completed in the proper format and does not look attractive and neat.
Required Elements Included **x2**	-Main character's job -Main character's passion -Identify THREE thoughtful, specific character traits with evidence -What barriers did the character face to reach their success -Outline steps taken in their research/process (5 Total)	4/5 elements present	3/5 elements present	Less than 3 three elements present

---

[16] Rubric designed by Mrs. Meghan Chakhtoura.

Constellations

As I shared through the bluebird's project, there are three levels of STEM integrations: "multidisciplinary or thematic integration, interdisciplinary integration, and transdisciplinary integration" (Vazquez et al. 59). The Constellation STEM project has characteristics of multidisciplinary and interdisciplinary approaches. The project has elements of the interdisciplinary approach because disciplines like science, math, technology, and art are combined to design the physical model of the constellation. Students need knowledge in electrical circuits and constellation, but they also need to use math skills to draw the constellation at the proper scale. It goes beyond science and math since students must be familiar with Chibitronics© technology[17] as well as being able to create the galaxy background in art class. Additional math skills are needed to write the distance between stars in scientific notation and to calculate the amount of copper adhesive tape needed. The multidisciplinary or thematic integration connects disciplines around the common theme "constellations." After our cluster meeting, we decided to include some extra activities on the constellation unit that were relevant to the theme and enhanced the STEM integration. Students read a story book on mythology in literature class, wrote a research paper on the chosen constellation in language art class, reflected on God's creation during Religion class, built quadrants to determine the latitude of Jacksonville in social studies class, and then closed the activity with a field trip to MOSH Planetarium. This field trip added relevance to the project since we were part of the planetarium program "The Sky in Jacksonville." After embarking on this journey across the stars and constellations, they left with a greater appreciation of our wonderful universe and the desire to explore constellations with their families.

The constellation project offered the opportunity to observe, describe, and explain the natural world. It was the perfect framework for the whole-child approach within the richness of the content, the opportunity to integrate disciplines, and the importance of collaboration and teamwork. It also offered the opportunity to emphasize our Catholic identity as students explore our ever-expanding universe with awe and reverence.

STEM Project Description: Students began this integration with a Stargazing Night where they were able to identify stars in the sky and discuss constellations with the North-East Florida Astronomical Society (NEFAS) members. The Engineering Design Process was introduced at the beginning of the school year from the *Science Fusion Program ~ Module: Engineering & Science*. After choosing the constellation and discussing constraints (size of constellation and maximum number of lights emitting diodes (LED)), we introduced the concepts of scaling, ratios, proportions, and scientific notation for scaling and stars' distances purposes. We introduced circuits and students sketched, designed, created, and tested the parallel circuit to light the dilated constellation.

---

[17] To learn more about Chibitronis©, visit https://chibitronis.com

Disciplines that were integrated are listed below:

- Science: electrical circuits
- Technology/Computer Science: use of Chibitronics©
- Engineering: sketch, create, test, and improve a solution.
- Mathematics: problem solving, scale factors, ratios, proportions, and scientific notation
- Language Art: research paper on constellations
- Literature: stories on mythology
- Religion: discussion on the Catholic identity "The Heavens are telling the Glory of God"
- Social Studies: building quadrants, determining the latitude in Jacksonville
- Art: constellation background
- Field trip to MOSH Planetarium "The Sky in Jacksonville"

Real world context: Students were asked to design an electrical circuit with Chibitronics© to light a constellation. Constellations are used to interpret sky patterns and for navigation purposes. Parallel circuits also have many applications since they are found in most devices: they are used to supply energy, for electrical wiring, and are used in most appliances.

Students' Task: Students used formulas to scale up the constellations and to express the stars' distance in scientific notation. Students used scientific concepts on electrical circuits to design the parallel circuit to light the chosen constellation. Students sketched, created, and tested their designed circuit. They also explained how they improved their design and made a written report on the chosen constellation (history, mythology, stars).

Assessment: Formal assessment using a rubric and informal formative assessment through different check points during the design and testing processes.

Target audience: 6th grade students (Astronomy) but can be implemented with 8th grade to apply electrical circuits content.

Instructional time: 6 hours.

To practice effective communication skills that allows students to build fluency and confidence, students presented their constellation models and research on the constellations during our annual "STEM Open House". The presentation included the history of the constellation, mythology, location of constellation, and stars & star patterns.

The *Jamestown- Yorktown Foundation Museums created a video about the 17th Century Maritime Celestial Navigation*[18] that led students to a hands-on activity to explore celestial navigation. They

---

[18] This activity was facilitated during Social Studies class by Mr. Michael Kavanagh. To learn more about this activity, go to https://youtu.be/5DO1wygRtPy?

built a device to measure latitude with interesting connections in the engineering field and the advancement on navigational techniques.

The following documents about the Constellation STEM project have been included:

- Diagram of the Interdisciplinary Approach
- Constellation Project Description
- Constellation Background Description
- Constellation Prototype Rubric ~ 6th Grade (The rubric includes galaxy background, transferring of constellation, parallel circuit layout, and the drawing of the constellation. It does not include the math component on scaling the constellation since this content is not part of the sixth-grade math curriculum.)
- Constellation Essay Rubric
- Photographs of two of the Constellation Prototypes

Interdisciplinary Plan to Light-Up a Dilated Constellation

Catholic Identity: The Heavens are telling the Glory of God.

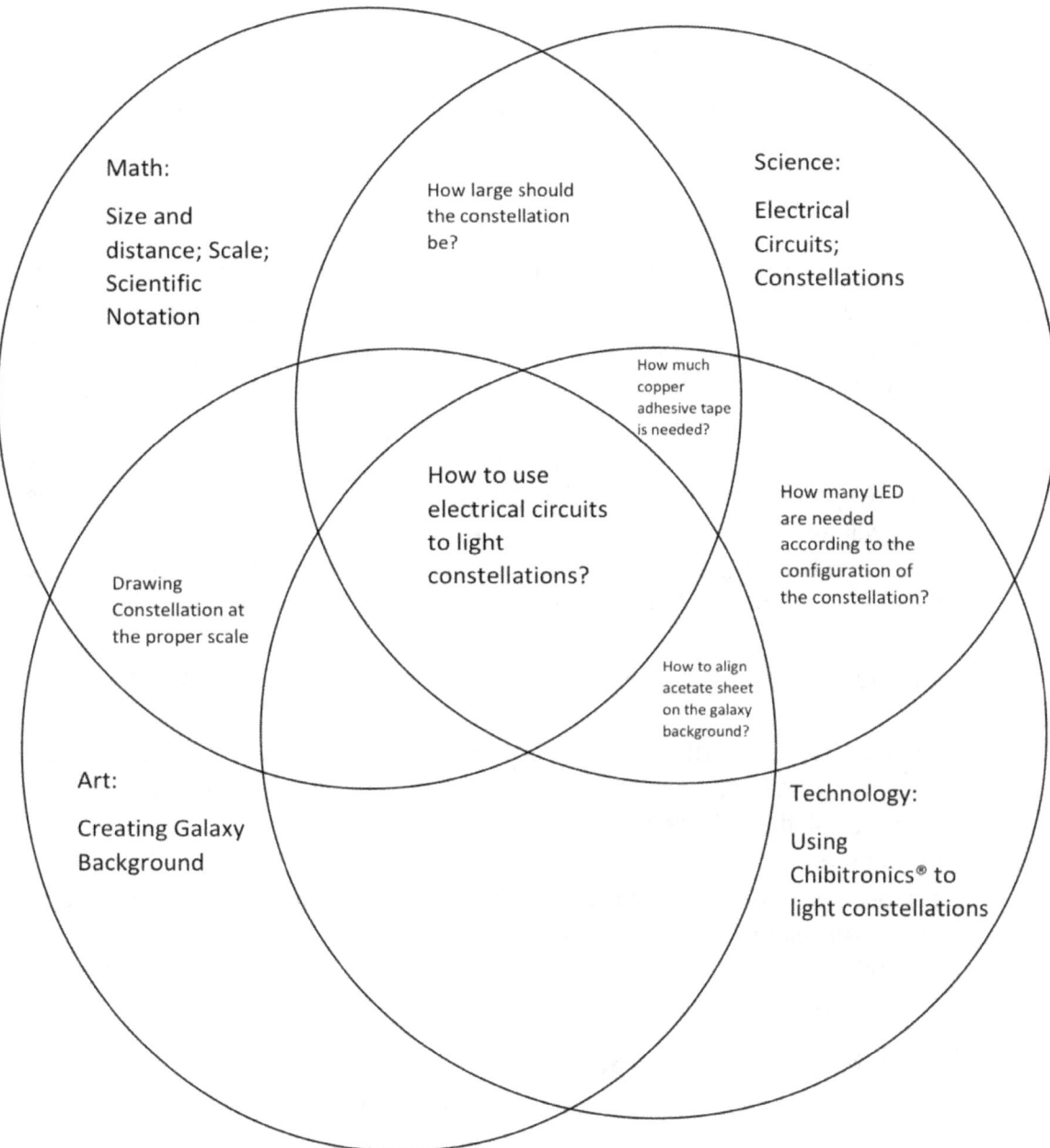

Additional Disciplines in the STEM Integration:
Religion: Reflection on God's Creation, English: Research Paper on Chosen Constellation
Literature: Story Book on Mythology, Field Trip to MOSH Planetarium ~ The Sky in Jacksonville

Parallel Circuits Constellations[19]

ASK: How can we design a parallel circuit to add lights to our dilated constellation?

IMAGINE: What do we want to light up?

1. Decide which team member's galaxy background you are going to use. Keep in mind the orientation of the constellation on the background.
2. Decide which stars (no more than 6) you want to light up. Make sure that each star you pick has room on either side of it for the adhesive copper tape.

PLAN: Sketch the parallel circuit on the circuit board to connect battery to LED stickers

3. Get a piece of white 8.5 x 11 heavy weight cardstock piece (circuit board).
4. Lay your galaxy background on top of the circuit board and line it up perfectly. Hold it still and mark a black dot through the holes with a pencil so you can see where you are going to place your LED stickers on the circuit board.
5. Do your best to place the battery compartment on the circuit board in the right location where you are going to want to push to light up the circuit. Do not get it too close to any edge or you will not have room for your copper tape.
6. Remove the tape backing off the battery compartment and stick it to the circuit board with the fold towards the bottom. Make sure it is about 1" from the bottom and 1" from the right edge.
7. Mark which side of the battery compartment will be positive (+) and which will be negative (-).
8. With a pencil draw a straight pencil line from the negative sign up to your first LED mark.
9. Draw a straight pencil line from the positive side of the battery compartment, then up to the other side of all the LED marks.
10. Once your lines are done, stop.
11. Use a ruler and do an estimate of how much copper tape you are going to need and write it on a scrap piece of paper.
12. Teacher Check Point – have your teacher check your work at this point. You will receive the rest of the supplies you will need to finish the circuit.

CREATE: Lay down the parallel circuit and LEDs

13. Peel off the adhesive on the copper tape only as you go, not too far ahead of what you are laying down or it will get tangled and twisted and make a mess.
14. When you get to turn, stop – fold the tape back in the opposite direction that you are going to go, press down slightly, and then fold it back in the direction you want to go forward.
15. For the positive line, start on the inside of the top folding piece of the battery compartment.
16. Bring the copper tape over the edge to the front, down the ¼ "depth and onto the flat surface of the circuit board.

---

[19] Mrs. Corrie Leas created this project description after math, science, and technology disciplines were discussed by cluster teachers.

17. Proceed as you did for the negative side remembering to turn the copper tape correctly and make sure to keep the two lines of tape apart, but no more than ¼ ", when you pass by a LED mark.
18. Use the back of your fingernail to smooth all the copper tape lines down as much as possible.
19. Carefully peel the sticky backing off only one LED sticker, try not to touch the back with your fingers, and place it on one LED mark with the negative side (pointy side) touching the negative copper tape strip and the positive side (wide side) touching the positive copper tape. Stop before doing the second one!
20. Place the battery in the compartment with the negative side facing down and the positive side up.
21. Gently press the top white piece of the battery compartment to the battery and see if your first LED lights up.

IMPROVE / REDESIGN: Fix problems with the circuit

22. If the first LED does not light up, then check your circuit:
    a. Is the LED sticker touching both lines of the copper tape completely? Press down on the LED sticker or shift it gently if necessary.
    b. If the LED sticker cannot reach both lines, then the lines are too far apart you might need to lay a small piece of copper tape next to the one line to make it closer together.
    c. Is the battery in the compartment in the correct direction?
    d. Does the copper tape in the battery compartment make contact with each side of the battery when you close the compartment?
    e. Is there a break in the copper tape anywhere? Is it pressed down firmly?
    f. If you cannot figure it out, ask a teacher for help.
23. If it does light up correctly, then stick on the next LED sticker over the next LED mark and check to make sure it lights up.
24. Use small pieces of double-layered foam tape to build up tightly around the battery in the battery compartment to keep the battery from moving around. Do not remove the green paper from the top layer of foam tape.
25. The foam tape can cross over the copper tape. It will not hurt the circuit.
26. Place some longer pieces of foam tape around the circuit board.

Scaling Constellations

(Scaling is just for eighth grade students. Sixth grade students will receive a constellation that fits in computer paper size.)

Overview

1. Students will use a small copy of the chosen constellation and scale it up to fit comfortably on an 8.5 x 11 piece of cardstock.
2. The students will identify the major stars in the constellation (at least 4) and research the distance from earth to each star in light years. Students will use ratios to convert the distance in light years to miles in proper scientific notation.

PART A - Dilate the figure of the constellation

Using the smaller figure of the constellation given to you on graph paper, determine a scale factor that will allow your constellation to fit comfortably on an 8 ½" x 11" sheet of cardstock with at least 1-inch blank margin all the way around the constellation. Follow the steps below to accomplish this.

3. Using the smaller figure (original) of the constellation, select a line on the grid paper that is either through or just to the left of the left-most star of your constellation. Draw a vertical line on that grid line on the graph paper. This will be your y-axis.

4. With the original figure, select a line on the graph paper that is through or just below the lowest star of your constellation. Draw a horizontal line on top of that grid line on the graph paper. This will be your x-axis.

5. Mark the origin (0,0), where the two lines intersect.

6. Count the number of squares wide and the number of squares high of the original constellation. Start counting at the origin in the bottom left corner.

    Original Width = _____ squares   Original Height = _____ squares

7. Determine if you want your dilated figure to be landscape or portrait.

8. On a new, clean sheet of 8 ½ by 11 graph paper (use the quarter inch side), turn the graph paper to either portrait or landscape orientation and draw a vertical line 4 squares in from the left edge and a horizontal line 4 squares up from the bottom. These will be the x and y axes of your dilated figure. Mark the origin in the bottom left corner where the two lines meet.

Educating the Whole Child in Catholic Education

9. Count the number of squares (and mark off every 2 or 5 squares) that you have in each direction while still leaving at least 4 squares from the edge. In portrait mode, this will give you about 0 to 25 on the x-axis and 0-35 on the y-axis. In landscape mode, this will give you 0 to 35 on the x-axis and 0 to 25 on the y-axis.

   Number of horizontal squares (width) available for dilated figure: _____

   Number of vertical squares (height) available for dilated figure: _____

10. To find the denominator of the scale factor, determine which dimension (width or height) of your original figure is larger. Choose the larger value and put it in the denominator of the scale factor.

11. The numerator of the scale factor is the corresponding dimension on the new graph paper. Determine the scale factor by dividing the number of squares for dilated figure by the number of squares in original figure.

12. Use a calculator to do the division and you should get a value > 1. This will be your scale factor. Since we are scaling the figure UP, the scale factor should be > 1. Round your scale factor DOWN to the nearest tenth.

    My Scale Factor is: _____

13. Now number each star in your ORIGINAL figure of the constellation with a small number starting at 1 and include every star.

14. Fill in the table with the coordinates of the stars in the original figure in the second column.

15. Before multiplying every point by the scale factor, check that the tallest point (star) and widest point (star) will fit on the new graph. If they don't, you will need to adjust your scale factor down until they do.

16. Now create the new coordinates for the dilated figure by multiplying each x and y by the scale factor and create a new ordered pair for the new point (star).

17. Plot the new coordinates (3rd column) on the dilated graph.

18. This new graph will then be used to mark the star positions on the galaxy background you will make in art class.

    Dilation of constellation to fit 8.5 x 11 using scale factor: _____

Star Number	Coordinates in Original Figure (x, y)	New Coordinates in Dilated Figure (x·scale factor, y·scale factor)
1		
2		
3		
4		
5		
6		
7		
8		
9		
10		
11		
12		
13		
14		
15		

PART B – Stars of the Constellation

19. Identify at least 4 stars in the constellation and put them in the table below:

Star Name	Light-Years from Earth	Miles from Earth in Scientific Notation

Educating the Whole Child in Catholic Education

20. Do research to find how far each star is from Earth in light-years. Put this in the table.

21. Using ratios and proportions, and the fact that 1 light year = 5.878 x 10¹² miles, convert each of the stars' distance from light years to miles. Express your answer in proper scientific notation. Show all your work. All work must be submitted with the project!

Galaxy Backgrounds with Distress Oxide Inks [20]

Students use an ink blotting technique using distress oxide inks, blotting sponges, card stock white paper, and vinyl gloves in this art process to create a galaxy style background for a continuing star constellation project.

Students use galactic references to see the beautiful color patterns that are "painted" across the galaxies within our universe. This will help in the creation of their space backgrounds.

To start, students create the first layer of the inking process by blotting bright sections of colored inks onto the paper, covering a large surface area of the paper. These colored sections will become nebulae or distant galaxies.

Next, students apply a second layer of inks using a rich black to create the outer space background. The black ink is blotted onto the paper using a sponge, as like the previous step. This time, the black ink should be concentrated around the perimeter of the paper and between the colored area patches.

Then, a lighter concentration of black ink is blotted on and around the colored patches to help tone the colors and give them a more distant appearance. Continue to blot until the paper is completely covered with the black ink, leaving small, muted patches of color feathered in with the black, giving the illusion of distant galaxies, nebulae, and milky ways.

Finally, a fine splatter technique with diluted white acrylic paint and a toothbrush is used to create the illusion of distant stars across the galaxies.

---

[20] The galaxy background was written by the art teacher, Mrs. Deanna Mohler. She facilitated this component of the project.

6th Constellation Project Rubric

Constellation Project	Exemplary: Consistent and independent. (4)	Accomplished: consistent with little to no teacher prompting. (3)	Developing: inconsistent and/or require teacher prompting. (2)	Beginning: experiences difficulty even with teacher prompting. (1)
Galaxy Background **Neatness** **Simulate galaxy background**	Generated a unique galaxy background. Constellation background is extremely neat and creative. Carefully considered the galaxy background to simulate it for the project.	One of the following is not included: Generated a unique galaxy background. Constellation background is extremely neat and creative. Carefully considered the galaxy background to simulate it for the project.	Two of the following are not included: Generated a unique galaxy background. Constellation background is extremely neat and creative. Carefully considered the galaxy background to simulate it for the project.	None of the following are included: Generated a unique galaxy background. Constellation background is extremely neat and creative. Carefully considered the galaxy background to simulate it for the project.
Transferring Constellation from paper to galaxy background	The constellation was carefully and accurately transferred from the paper to the galaxy background such that all the holes line up properly with the LEDs.	The constellation was transferred but had one error.	The constellation was transferred but two errors.	The constellation was transferred but with more than two errors.
Parallel Circuit layout	The positive and negative sides of the parallel circuit reach every star that is going to be lit. Intersecting circuit lines are taped to prevent short circuits. Battery compartment is thoughtfully located within the circuit. Copper tape is attached very well with few tears and all tears are fixed with patches. The circuit works correctly when the battery compartment is pushed.	1 or 2 of the following are not included: The positive and negative sides of the parallel circuit reach every star that is going to be lit. Intersecting circuit lines are taped to prevent short circuits. Battery compartment is thoughtfully located within the circuit. Copper tape is attached very well with few tears and all tears are fixed with patches. The circuit works correctly when the battery compartment is pushed.	3 or 4 of the following are not included: The positive and negative sides of the parallel circuit reach every star that is going to be lit. Intersecting circuit lines are taped to prevent short circuits. Battery compartment is thoughtfully located within the circuit. Copper tape is attached very well with few tears and all tears are fixed with patches. The circuit works correctly when the battery compartment is pushed.	None of the following are included: The positive and negative sides of the parallel circuit reach every star that is going to be lit. Intersecting circuit lines are taped to prevent short circuits. Battery compartment is thoughtfully located within the circuit. Copper tape is attached very well with few tears and all tears are fixed with patches. The circuit works correctly when the battery compartment is pushed.
Acetate Clear layer with drawing of constellation	The constellation is accurately drawn on the acetate. Orientation and scale of the constellation have been considered. The drawing overlays on the star pattern accurately.	One of the following is not included: The constellation is accurately drawn on the acetate. Orientation and scale of the constellation have been considered. The drawing overlays on the star pattern accurately.	Two of the following is not included: The constellation is accurately drawn on the acetate. Orientation and scale of the constellation have been considered. The drawing overlays on the star pattern accurately.	None of the following are included: The constellation is accurately drawn on the acetate. Orientation and scale of the constellation have been considered. The drawing overlays on the star pattern accurately.

Students' Names

_____

Eighth Grade ELA – Constellations Essay – Rubric [21]

History Students include factual information about constellation's history	_____/3
Discovery and naming of the constellation are explained	_____/2
Mythology Students offer at least one mythological account for the constellation	_____/2
Myth is effectively summarized	_____/3
Stars Students give examples of names of stars	_____/2
Students describe star patterns and scientific terminology is clear	_____/2
Scientific terminology is clear	_____/1
Conventions Spelling and Capitalization	_____/2
Punctuation	_____/2
Grammar and Usage	_____/2
Sentence variety and fluency	_____/2
Presentation Works cited are present and correct	_____/2
Essay is double-spaced	_____/1
Essay is stapled (if necessary)	_____/1
Total	_____/27
Total **Final Grade** _____ %	

---

[21] This rubric was designed by Mr. Michael Kavanagh.

Photographs of two of the Constellation Prototypes

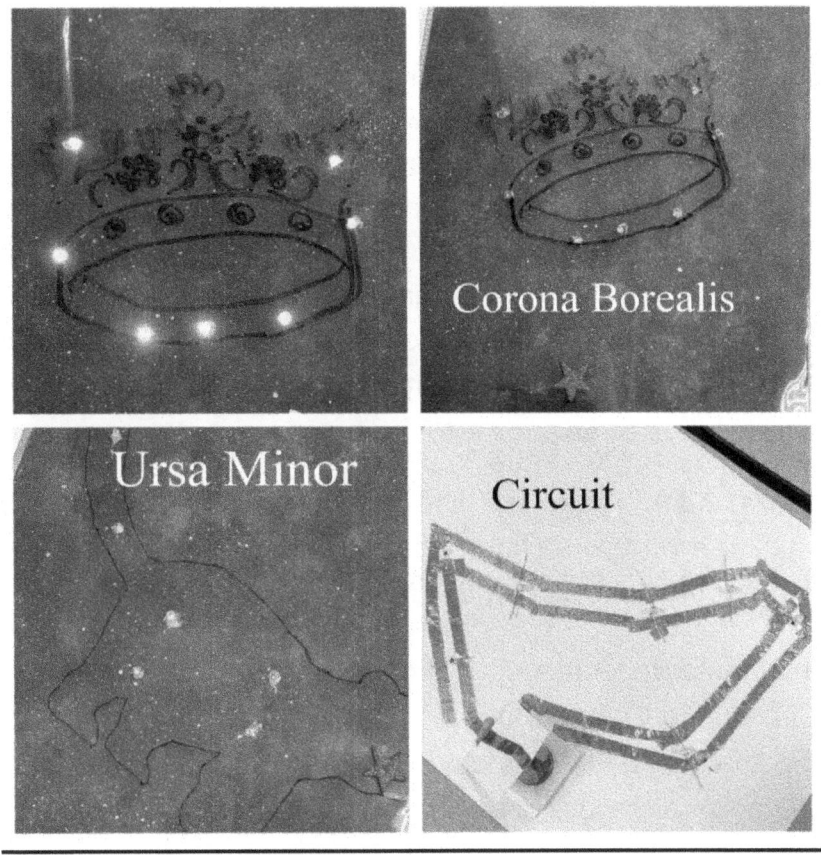

First Square: Lighted up Constellation

Second & Third Squares: Transferred Constellation on Acetate Clear Layer

Fourth Square: Parallel Circuit Layout on Circuit Board

The Free-Market Project

Students had to design a product that is not currently on the market. They created the prototype in Tinkercad[22], which is a free web app for 3D design, electronics, and coding. It allowed students to create a 3D printed model during STEM class. Seventh grade students presented their project during our annual "STEM Open House". Having the physical model and the digital design in Tinkercad gave students the opportunity to showcase their product in a more effective way. Students also designed a commercial for marketing purposes. This project represents a great model of integration of disciplines in a social studies project with a meaningful Catholic component. One of the seven themes of Catholic Social Teaching is the Dignity of Work and the Right of Workers. How a society functions has a direct impact upon the dignity of the human person and the ability to flourish in conjunction within community. Individuals have a right to engage in a shared life within society working towards the common good.

This project is a great example of how to design a project in a different discipline under the STEM umbrella. As you read this social studies project, you will find a focus on inventions and the use of new technologies for design and marketing purposes. This project contains relevant elements of a STEM approach:

- Involves collaboration
- Contains real-world context
- Incorporates a relevant topic
- Driven by a process to design a product
- Allows multiple solutions
- Incorporates the use of technology
- Addresses real-world problems

---

[22] To learn more about Tinkercad, visit https://www.tinkercad.com

The Free-Market Project[23]

You will be assigned to groups of two or three. Each group will be responsible for creating a product that is currently not on the market.

- It has to be a product that improves a person's life.
- It has to be a product that has not been created before.
- It has to be a realistic product that could be introduced in the next few years.
- It can be anything, from an electronic to a new kitchen tool. Be creative!

Stage 1: Idea and Conceptualization

Brainstorm ideas for a new product. Consider how realistic a product would be and whether it would make a positive impact on a person's life. You will need to submit two paragraphs (8 sentences each) describing 1) what your product does and 2) how it would improve a person's life. All products must be approved by teacher.

Stage 2: Design and Model

Once you have settled on a product, you will need to come up with a prototype model. You will work with your group during STEM Class to create a 3-D printed model of your product. Each group must print and submit a 3-D model of your product. The model does not have to be a full-size model.

Stage 3: Research costs of production (Production Cost Worksheet)

You will need to set a price for your product to sell on the market. In order to do that, you will need to have an understanding of how much it costs to produce your product. Each group must:

- create a list of all the materials you will need to build your product,
- research the prices of materials needed (be as realistic as you can),
- set a price for your product (be sure to factor in a profit for your business).

Keep in mind the law of supply-demand: the higher the price, the less people will be willing to buy it. You will need to submit a list of all the materials needed to make your product and their prices. Think of this as "showing your work" for how you determined the price for your product.

---

[23] Social studies project designed and facilitated by Mr. Daniel Conigliaro

Stage 4: Marketing

Now that you have a product and price, you will need to come up with a marketing strategy for your product. Who is your target audience for this product? Who would be most likely to buy this product? Be creative with your ad! Your marketing strategy must include each of the following:

- TV commercial (no more than 30 seconds)
- Flyers for the classroom
- "Billboard ad" (poster)

Stage 5: Selling your product

Now that you have advertised your product, it is time to bring it to market! You will have an opportunity to set up a market stand in which you will "sell" your product. Each group will need to create a tri-fold poster display for their market stand. The market stand must include:

- Images and words to describe your product
- Visual appeal

Everyone in the class will buy up to two items in the marketplace. At the end of the marketplace day, each group will need to submit a tally of how many goods they sold (this is the "revenue sheet"). The group that sells the most will receive one "Homework Pass" good for any homework assignment in Social Studies.

Rubric for Grading:

Product approval: 10 points

Product design/model: 20 points

Production costs research/worksheet: 20 points

Advertisement: 30 points

Market stand: 10 points

Revenue sheet: 10 points

TOTAL: 100 points

This assignment will be a project grade for Social Studies.

Chapter 5

Center for STEM Education ~ University of Notre Dame[24]

> "Let the Word of Christ dwell in you richly; teach
> and admonish one another in all wisdom."
> Colossians 3:16

STEM education provides a framework that recognizes the contributions that each student's identity can bring to the classroom. With this focus on identity, Catholic schools express their most distinctive mark, their Catholic character, through the intersection of faith and reason in a faith-based context. STEM education centers on the whole child with a strong foundation on skill development and the acquisition of knowledge within an interdisciplinary approach. This STEM framework aligns with the goal of educating the child to form future leaders to solve problems, taking into consideration the ethical and moral implications of their decisions. When our faith and values are central to this interdisciplinary approach, Catholic teachers are collaborating with God's grace to form people for Heaven. This is how we make a valuable distinction to the whole child to aid in the formation of saints for God's Kingdom. The Center for STEM Education at the University of Notre Dame was a God-given opportunity to incorporate the principles of STEM education with academic rigor and within a faith-based setting.

It is hard to put into words the impact the Notre Dame STEM Teaching Fellowship had in my personal, professional, and spiritual life. This professional development program connected me with STEM educators across the country and provided me with the tools to enhance STEM education in the classroom. It consists in three main modules:

- Module 1~ *STEM Integration* where fellows are exposed to STEM units as students and as teachers. It gives participants the opportunity to reflect on the integration from different points of view as they go over core disciplinary concepts and practices.
- Module 2 ~ *Core Instructional Practices* where fellows acquire tools to bring out students' insight in a more effective way. It includes sharing ideas, facilitate conversations and discussions in the classroom, providing feedback in an

---

[24] To learn more about the Notre Dame STEM Teaching Fellowship, go to the link below https://stemeducation.nd.edu/
It will give you information about their mission & vision, core belief, faculty & staff, research, programs, and teaching fellows.

assertive and effective way during instruction and assessments, and using phenomena from the natural world to find answers through a claim-evidence approach.
- Module 3 ~ *STEM Impact Plan* so teachers could return to their schools with a STEM Blueprint Plan to implement. The plan is based on the school context and culture to provide a relevant and realistic plan for each participating school.

This long-term professional development was piloted in three years with a focus on instructional practices for science, mathematics, and engineering & STEM integration. Some of the most relevant practices in my professional journey through this professional development were:

- Modeling Science Phenomena
- Assessments in STEM Integrations
- Embracing the Uniqueness of the Human Person
- STEM Blog ~ Center for STEM Education

<u>Modeling Science Phenomena</u>[25]: using different methods to generate class discussions on core science ideas through raising questions is an essential task for teachers. The use of videos to observe phenomena is a meaningful method with several advantages:

- Students try to make sense of what they are observing instead of listening to a teacher's lecture.
- You can target the visual learners.
- This is a form of informal assessment where teachers can check prior knowledge.
- Because it is an informal assessment, students work without stress and feel more comfortable to cooperate in class discussions.
- It creates a positive classroom culture.

Effective teachers welcome and appreciate good questions. Being open to questions is important, but developing the art of asking good questions is also crucial to drive students to the explanation of phenomena from the natural world that are justified with evidence. Teachers' questions must serve the purpose of sharing ideas and listening critically to other students' ideas. This way students may arrive at answers on their own, which are the product of a collaborative effort.

---

[25] Modeling Science Phenomena through videos was facilitated by Dr. Matthew Kloser, Core Instructional Practices leader at the Center for STEM Education ~ University of Notre Dame.

One example to illustrate the use of videos and productive talking is presented below:

Spoiled Milk video: you can find many options online that will fit this purpose, the one presented in this unit is no longer available. The video was introduced after teaching chemical reactions and preparing to introduce bases, acids, and salts. The STEM challenge at that time was about how to solve a food storage problem, so it gave us the space to continue talking about processes like pasteurization. This video can also be useful to introduce chemical reactions and the signs that a chemical reaction has taken place.

After presenting the video, ask questions about the phenomena as you try to generate a class discussion that can lead to an explanation of what they are observing. An example is provided to illustrate the idea of driving students to an explanation rather than lecturing them about the natural world's phenomena:

Teacher: What do you see? Please write your observations before we start our discussion.

Student #1: The man spilled the milk.

Teacher: Why do you think the man spilled the milk?

Student # 2: It may be that the milk had a bad taste.

Student # 3: The man did not smell the milk.

Teacher: Why do you think the man had to smell the milk?

Student # 3: Because when the milk has been in the refrigerator for too long, it smells bad.

Student #4: In addition of smelling bad, it also tastes sour.

Teacher: What would be the reason of the changes in odor and taste? Please discuss this question in your small groups before we go back to the whole class discussion.

Group # 3: The change in taste is telling us the milk is spoiled. It has happened to some of us when we drink the milk without checking the expiration date.

Group # 5: Change in odor is one of the signs that a chemical reaction has taken place.

Teacher: Let's think about the ideas you have shared about the sour taste and the change in odor in light of a chemical reaction. What happens at the invisible level that the milk has a bad taste and has a change in odor?

Student # 4: It seems that a chemical reaction has taken place.

Teacher: Do we know for sure that a chemical reaction has taken place?

Group # 7: We discussed it in our small group and a new substance needs to be formed in order to conclude that a chemical reaction has taken place.

Teacher: Do you think the new substance, if any, has changed the flavor of the milk?

Group # 1: The formation of a new substance may be the reason of the bad taste.

Teacher: How are you thinking about your responses now? Please answer the question by getting together in your small groups and drawing a model that can explain what is happening at the microscopic level with that milk. How do you know a new substance has been formed? Can someone explain that in their own words?

During the discussion, writing the evidence of the students' responses is vital to encourage the use of common language and to empower students' thinking. During the small group interaction, it is important to circulate and check if students are using common language. This is also a good moment to start thinking about which models/drawings you will select for discussion to avoid redundancies. In the case of the spoiled milk discussion, drawings can provide some relevant discussion points about the presence of lactose in milk, expiration dates, and the pasteurization process. Further discussion and additional research would lead to a more detailed explanation about the changes in the milk due to a chemical reaction. During that chemical reaction, bonds are broken to form new substances with new properties. Milk contains a sugar called lactose and due to certain bacteria, lactose and oxygen combine to form lactic acid, which explains the spoiled milk taste. This is an effective way to make students talk about what is happening at the visible and invisible levels, and even more relevant, to draw conclusions about how the activity at the invisible level affects the visible level. "Students can be provided with time toward the end of the lesson to modify their representations based on what they have learned in the whole-class discussion, thereby finishing the lesson with a complete and accurate representation/explanation of the scientific phenomenon they are studying." (Cartier et al. 88)

Further discussions were related to the importance of checking the expiration date of products and the value of using science to make smart decision. These days, students know about many challenges found on social platforms like the "spoiled milk challenge." Students were able to articulate, with confidence, the science behind spoiled milk and the reasons they must not take part in these social media trends. One interesting extension activity is analyzing the risks and benefits of the pasteurization process. Milk is pasteurized by the process of heating it at about 150°F to kill bacteria. A specialist in food engineering came to speak to the class, providing meaningful insights to this process as well as other processes like packaging, manufacturing, and safety when handling milk and other products.

"Knowing how students are making sense of scientific phenomena is critically important for teachers who want to help students become more independent in their thinking about and evaluating scientific knowledge. Focused talk that can bring their thinking to the surface is therefore an essential ingredient of expert pedagogy" (Cartier et al. 89). The use of videos, experiments, demonstrations, and images engage students in meaningful discussions and allow students to make sense of phenomena in a most relevant and impactful way. Some good practices when generating productive discussion are:

- Avoid good/bad or right/wrong when providing feedback.
- Showcase evidence of students' thinking: Predict-Observe-Explain (POE) charts are a great option as well as charts that can contrast ideas, evidence, and relation to the phenomena in discussion.
- Grasping ideas.
- Noticing the cause and effect of events.
- Avoid the urgency to get the right answer.
- Provide concrete examples for a class discussion that can foster the right classroom culture.
- Empower students to participate: walking during small class discussion is a great opportunity to ask students to shout out an idea you have heard.
- Present relevant and open-ended questions.
- Rephrase students' ideas.
- Promote student-to-student communication.

Assessments in STEM Integration ~ Pendulum System ~ STEM Lab[26]

When students are assessed formatively throughout a STEM integration, the use of a summative assessment is imperative. These task-based summative assessments evaluate content and skills that are taught and addressed through STEM integrations. They also offer the opportunity to measure holistic performance with rubrics that contains level descriptors, categories, scales, and weighting. The categories break down the unit goal, but those that are essential to the unit goal are multiplied by a factor.

I have combined in the STEM lab "Pendulum System", some effective ways to assess students through this STEM integration. The goals of this STEM integration were:

- Designing an assessment after the laboratory activity where students could be individually assessed. This lab activity is part of the **F**ull **O**ption **S**cience **S**ystem FOSS[27] program. The module's name is **Variables** and the system chosen is **Pendulum.** Students conducted a controlled experiment to determine which variables (mass of the bob, release position, and length of the pendulum) affect the number of cycles a pendulum will complete in 15 seconds.
- Using "snowballs" to check students' understanding on the effect of the variables in the number of swings in 15 seconds. This is also a way to find out in which directions their hypotheses will go.
- Designing a rubric to measure holistic performance.

Students' Learning Goals:

- Design and present a controlled experiment with a testable hypothesis.
- Analyze and interpret data to determine similarities and differences in findings.
- Construct a scientific explanation based on valid reliable evidence obtained from data and research.

Snowball Formative Assessment: Class discussion led to testing the variables: length, mass of the bob, and release position. For this STEM lab, I used it before students wrote their hypotheses, but it has multiple uses. In this particular case, it allowed me to find out in which direction students' hypotheses will go. The advantage of this type of assessment is that it does not create competition, but a community of learning. An additional advantage is that it is an easy and quick way to check students' understanding before moving forward. The "Snowball Formative Assessment" consists of asking students to answer a question on a piece of paper (providing index cards or cut computer paper

---

[26] Assessments in STEM integration was facilitated by Dr. Gina Svarovsky, STEM Integration leader. The Snowball Formative Assessment was part of the module on Core Instructional Practices facilitated by Dr. Matthew Kloser.

[27] To learn more about this program, visit https://foss.lawrencehallofscience.org

ensures answers are anonymous.) They must crumple it up into a ball and on the teacher's direction, students throw the paper to the front of the classroom. They will then pick one paper and share the answer with the class. If you want to expand on this activity, you can ask students to explain their answer. Students really enjoy this type of activity where they need to move around the classroom. It can be unstructured at the beginning, but it will become part of your class culture quickly.

Snowball Question ~ Pendulum System:

We will be testing the variables: release position, mass of the bob, and length of the pendulum in our investigation on pendulum systems. Which variable will make a difference in the number of swings?

   a. The three variables
   b. None of the variables
   c. Length
   d. Mass and Release Position

Documents for Assessments ~ Pendulum System:

The following documents were used to assess the Pendulum System ~ STEM Lab:

- Variables Chart
- Claims-Evidence-Justification Chart
- Assessing Science Ideas and Skills on Scientific Method ~ Formative Assessment for Lab Reports
- Individual Assessment ~ Summative Assessment
- Holistic Rubric Template

Variables Chart ~ Pendulum System

Students receive this worksheet after discussing the variables to be tested and the protocol/procedure to follow to perform the experiment.

	Hypothesis	# of Swings	Conclusion
Standard Pendulum System: Length: 38 cm Mass: 1 penny Release position: Straight out to the side (90°) Time: 15 seconds			
Changing the Variable "Release Position": Length: 38 cm Mass: 1 penny Release position: 45° Time: 15 seconds			
Changing the variable Mass: Length: 38 cm Mass: 2 pennies Release position: Straight out to the side (90°) Time: 15 seconds			
Changing the variable Length: Length: different length value in each group Mass: 1 penny Release position: Straight out to the side (90°) Time: 15 seconds			

A Claims-Evidence-Justification Chart is a useful tool to develop the scientific practice of argumentation. "Claims include the ideas, conjectures, or explanations that may provide a sufficient answer for a scientific question. Evidence includes gathered and analyzed data that can be used to support the claim. Justifications not only link the evidence to the claim, but they also provide a reasoning for why the evidence is important and how the evidence makes sense in the light of existing concepts and assumptions." (Kloser and Grathwol 16). A valuable recommendation from the authors is to provide a blank table with only the header titles.

The chart below is just an example of what students have done after experimentation. Students' charts may vary and include less or more information.

Claims-Evidence-Justification Chart ~ Pendulum System

Claims	Summarized Evidence	Justification
A pendulum is a weight suspended from a pivot so that it can swing back and forth. Pendulums have several uses and applications.	A pendulum can be created using a string and a weight. The weight can be designed by connecting a paper clip in one of the ends of the string and adding pennies in the paper clip to test the variable "mass of the bob." You can see pendulums in grandfather clocks. Every time you sit in a playground swing, you are experiencing the back-and-forth motion of a pendulum.	A pendulum is an object that swings freely consisting of a mass that sways from one end of a string or rod. Pendulums have several uses and engineering applications. They are found in amusement park rides and are part of the mechanisms of seismometers and clocks.
The motion of a pendulum may be affected by a number of factors (variables), such as its length, mass of the bob, and release position. Gravity and inertia also have an effect on pendulum systems.	As the pendulum swung back and forth, we tested three variables: length of the pendulum, mass of the bob, and release position (independent variables). A controlled experiment was designed testing one variable at a time. The dependent variable (number of swings in 15 seconds) just changed when the length of the pendulum was changed. As the length increased, the number of swings in 15 seconds decreased. For a 120 cm – pendulum, the number of swings in 15 seconds was 7, while a pendulum of 33 cm (control group) provided 13 swings in 15 seconds. The number of swings in 15 seconds was always the same when the mass of the bob and the release position were changed. The control group was a 13-cm pendulum with a mass of one penny and the release position was straight out to the side (90°) Gravity brings you back to the ground when you get too high in a playground swing, but inertia is what keep you moving back and forth.	Several factors affect the number of swings of a pendulum. The pendulum stays in motion due to inertia. The First Law of Newton states that an object in motion will continue in motion, and an object at rest will stay at rest, unless acted upon by an outside force. It explains why there is an outside force when you are swinging in a playground swing. Gravity is the force pulling the mass down. Gravity pulls you down while inertia keeps you moving until friction does its job. The motion of a pendulum was mathematically described by Galileo ~ Period = $2\pi / \sqrt{\text{length of pendulum/gravity}}$ who found out through experimentation and careful observations that the period of the pendulum depends on the length of the pendulum. This relationship has been used for centuries to adjust grandfather clocks. The designed controlled experiment proved that period of pendulums does not depend on the mass of the bob or release position, but the length of the pendulum.

After experimentation, each group wrote a lab report with the following information:

- Purpose
- Hypothesis
- Dependent and Independent Variables
- Constants or Controls
- Quantitative Measurements
- Experimental and Control Groups
- Materials & Procedures
- Chart & Graph
- Conclusion Statement

This small group report represents a formative assessment. No grades were given but feedback was provided. The advantage of this formative assessment is to prepare students for a summative assessment to evaluate their knowledge on controlled experiments, analysis and data interpretation, and constructing scientific explanation. The guideline below was used to provide feedback for the formative assessment. Students responded very well to getting feedback without having to worry about a grade during the group work. This is about creating a classroom culture where students are seeking understanding. Students wanted to fix the answers to feedback questions even though their work was not being graded. It is the right class culture with the great advantage of students feeling more confident by the time they take the individual assessment.

Assessing Science Ideas and Skills on Scientific Method

Formative Assessment for Lab Reports

Skills/Ideas	4	3	2	1	Description/Expectations
Writing a Hypothesis					Carefully constructed to answer the question: Does the variables (length, release position, and mass of the bob) affect the # of swings? Testable
Identifying Dependent & Independent Variables					What do you know before doing the experiment? / Factor you are manipulating. What do you know after doing the experiment?
Identifying Quantitative Measurement					Unit to measure the dependent variable.
Identifying Controls/Constants					Factors that must be controlled in order to test one variable at a time.
Designing/Interpreting Scatter Plot Graphs					Titles for chart and graph. Names and dates. Line of the Best Fit.
Writing Experimental Procedure					Step by Step procedure for others to be able to replicate the experiment.
Writing Conclusions					Is the hypothesis approved by the results? Use of evidence to justify the conclusion statement.

As stated before, the summative assessment measures students on:

- How to design a controlled experiment with a testable hypothesis
- How to analyze and interpret data to determine the similarity in findings
- Construct a scientific explanation based on valid reliable evidence obtained from data and research

Concepts and skills that are intrinsically part of the nature of science are being assessed. Students had the opportunity to continue developing those skills. In fact, to discuss the next chapter, students designed a controlled experiment to determine the effect of friction on motion. Every chapter has the potential to provide the opportunity to practice these important scientific skills.

When we have provided formative assessment throughout the educational experience, a summative assessment is necessary. The summative assessment and the holistic rubric to measure students with a holistic approach have been included. The holistic rubric provides structure and clear expectations as well as an ordered arrangement to grade students' performance. It contains:

- Level descriptors.
- Categories.
- Weight: The categories essential to this STEM lab are multiplied by two. Professionalism is included in the rubric, but it is weighted differently (multiplied by 0.5).
- Scale: Scale goes from one to four. Zero is not used because zero is for someone who did not do the assignment.
- A grading scale was used (see grading scale below the rubric) since it is necessary when using holistic rubrics.

It is important to look for participation and ask for evidence. It is essential to check if students know how to interpret the data, which allow students to use it as evidence. Some basic questions to ask are:

- What are the dependent and independent variables?
- How do we express the variables on a graph?
- What does a data point represent?
- What is the relationship between the variables?
- What pattern(s) do you notice?
- What does the trend say about the relationship between the number of swings in fifteen seconds and the variables length of the pendulum, mass of the bob, and release position?
- Ask students to tell you a story about the pendulum system.
- Relate data to experiences and science ideas: some students relate data to a playground swing while others go to the context of amusement park rides. Other connections may be related to a thurible that swings freely back and forth to dispense incense smoke during Mass or other engineered objects like clocks and seismometers. It depends on the identity of each student, which will be the next topic of discussion.

Individual Assessment / Summative Assessment

Answer the following questions independently and individually:

1. Eighth grade students were experimenting with pendulums to understand how they work. Each team made a swinger of a different length, and they shared the results in the following table:

   Number of swings in 15 seconds of pendulums with different lengths.

Length in cm.	No. of Swings
13	20
20	15
38	12
70	9
90	8
120	7
170	6
200	5

   a. Is there a relationship between the length of the pendulum and the number of swings in 15 seconds? Use evidence from the experimental results to justify your answer.

2. Use the line graph below to answer the following questions:
   a. How can you predict how many swings a 150-cm pendulum will make in 15 seconds?
   b. What is the independent variable in this experiment? How do you know?
   c. What is the dependent variable in this experiment? How do you know?

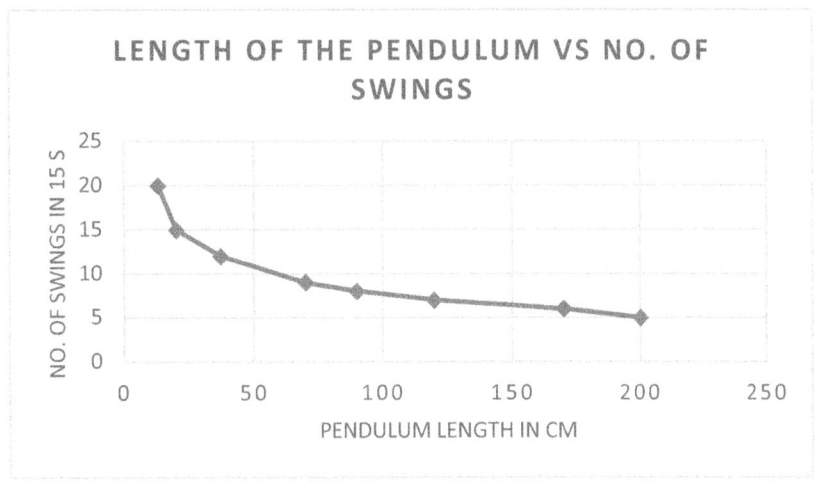

Educating the Whole Child in Catholic Education

3. One of the members of your team wanted to know what would happen with the number of swings if he/she changed the mass of the bob and the material of the string. Your classmate used fishing line instead of yarn to build the pendulum and he/she also used a washer at the end for the pendulum bob instead of a penny. Then he/she counted how many times his pendulum swung back and forth in 15 seconds. Do you think he/she has done a good job on controlling the variables? Explain your answer.

4. Three girls wanted to see who could throw a baseball the farthest. They traced a line on the playground that served as a reference point for them to throw the ball. The first girl stood still to throw the ball. The second girl ran to the reference point to throw the ball. The third one jumped on the reference point to throw the ball. The second girl threw the ball the farthest.
    a. Which variables did the girls control in the contest? How do you know?
    b. How would you change the procedure of the contest to better determine who throws the farthest? Explain your procedure step by step so others can replicate it.

Holistic Rubric Template: Pendulum System

	Exemplary 4	**Proficient** 3	**Basic** 2	**Emerging** 1
Designing Controlled Experiment x2	Experimental design is logical, detailed, workable. Collect measurable results using multiple trials. Variables and constants are identified and controlled.	Experimental design is generally logical, detailed, workable. Collect measurable results using multiple trials. Variables and constants are identified and controlled.	Experimental design is somewhat logical, detailed, workable. Collect measurable results using multiple trials. Variables and constants are identified and controlled.	Experimental design is not logical, detailed, workable. Collect measurable results using multiple trials. Variables and constants are not identified and controlled.
Analyzing and Interpreting Data	Interpretation of results is accurate and explains the differences and similarities in findings.	Interpretation of results is generally accurate and explains the differences and similarities in findings.	Interpretation of results is somewhat accurate and explains the differences and similarities in findings.	Interpretation of results is not accurate and does not explain the differences and similarities in findings.
Construct a scientific explanation x2	Student explanation shows an advanced understanding of the scientific method and it is based on evidence obtained from the experiment or research.	Student explanation shows a proficient understanding of the scientific method and it is based on evidence obtained from the experiment or research.	Student explanation shows a basic understanding of the scientific method and it is based on evidence obtained from the experiment or research.	Student explanation shows little to no understanding of the scientific method and it is not based on evidence obtained from the experiment or research.
Professionalism x0.5	The assignment is completed neatly. Exceptionally strong control of standard conventions of writing.	The assignment is completed neatly. Strong control of conventions; errors are few and minor.	The assignment is mostly neat. Limited control of conventions; frequent errors do not interfere with understanding.	The assignment is sloppy. Frequent significant errors may impede readability.

Grading Scale: 20-22 A, 19-18 A-, 15-17 B+, 13-14 B, 11-12 B-, 10 C+, 9 C, etc.

Embracing the Uniqueness of the Human Person[28]:

Even though we have an indelible mark of identity as disciples of Christ, each individual has a combination of traits that set them apart from anyone else, making them unique. Every person experiences different environments, relationships, family life, and events that shape their personality. It may translate into different interests, hobbies, lifestyles, priorities, customs, and ways of learning. As teachers, recognizing the uniqueness of each person is vital to fostering empathy and respectful interactions in the classroom as students develop a sense of belonging to thrive in academic and personal matters.

Since learning and identity are in a mutual relationship, considering students' identities would make learning relevant to each classroom. Identity includes ability, faith traditions, nationality, socioeconomic status, among other aspects. Identities may or may not be intentional, but in both cases, identities must be considered to provide relevant learning opportunities for all. Some aspects to be considered:

- Provide a variety of participation structures: calling on students, asking for volunteers, etc.
- Provide different types of assessments.
- Provide a variety of classroom activities geared towards different learning styles.
- Build upon students' cultural practices but also build upon their understanding of the primary culture in the classroom.
- Create a more diverse scientific discourse, including the influence of scientific issues on populations of different backgrounds.
- Celebrate and promote community as you help students to consider a range of perspectives.
- Position students as helpful humans as they become a force for good through the development of ethical solutions and the development of a critical and empathetic conscience.
- Relate content to personal experience.
- Make an intentional but sincere effort to make students feel loved and accepted.
- Emphasize our Catholic faith, which is, in most cases, a common identity in the classroom.
- Invite speakers from different backgrounds.
- Generate class discussions about the biographies of various scientists that are part of your science curriculum.

---

[28] Topics on Culturally Responsive Teaching were facilitated by Dr. Tia Madkins.

- Bring your own sense of who you are to the classroom and show how your identity affects your perspective about scientific and nonscientific issues. Modeling is always the best practice.

Students bring their whole selves to the classroom and part of our job is to discover who God created each of them to be. God has already imprinted his image in every human being. God has given us a unique personality as well as strengths and abilities; and exploring the characteristics that make each child unique is essential to the teacher-minister. While individuality is of value, the unity shared as many members of the Body of Christ is of greater importance to emphasize in this section. "... that there may be no dissension within the body, but the members may have the same care for one another." 1 Corinthians 12:25.

The use of identity maps is an interesting proposal. "An identity map is a tool to explore the multifaceted nature of what makes each of us unique" (Blue and Redick 49). An identity map looks like a diagram where students write their name in the center and then they share about their characteristics, interests, and strengths to show their nature and uniqueness. To help students feel more comfortable sharing about their identities, share an identity map of yourself and include the identity map of some known scientists. Creating identity maps at the beginning of the school year promotes a climate of mutual respect. The maps have the additional advantages of helping teachers to create a classroom community with students who are committed to promote differences as well as being a source of valuable information for the teacher to start learning about each student.

We used diapers during our culturally responsive teaching lesson at the Center for STEM Education at the University of Notre Dame. I kept the diaper theme, but with a different approach. Analyzing inventions was a common practice in my classroom, so I asked students to perform a risk/benefit analysis of the invention of disposable diapers including the evaluation of environmental impacts and the associated costs. Students' answers were as varied as their personal experiences. For some students, this invention was very familiar since they had younger siblings, while others shared that disposable diapers were not a common practice in their parents' times or countries of origin. We discussed advantages like the reduction of diaper rash and keeping babies dry thanks to the absorbent materials they are made of. However, the impact on the environment was also discussed, as these diapers are not reusable like the cloth diapers that are more natural. This led us to use more scientific terms and concepts since there is a material inside the diapers known as sodium polyacrylate with interesting properties and uses. As always, there is not a perfect solution to the dilemmas we discuss in the classroom, but a risk/benefit analysis provides people with the tools to make an informed decision where scientific facts, perspectives and cultural beliefs are all considered.

A student whose parents are from Dominican Republic went beyond this discussion. This student developed a science fair project in botany motivated by the issues that arise when crops are grown in vulnerable areas of the island. To promote forestation, the student proposed the use of polymers from diapers to engineer a soil that can maintain the humidity to mitigate water loss in places with dry climates. This is a great example of how identity impacts decisions and plans of action. The student placed himself as a "helpful human" with the desire to be a force for good to address a problem in the community.

Another example of the value of this type of approach was with the issue of single-use plastic bottles and its implications on the environment under the umbrella of our role as stewards of all God has created. Students had the opportunity to see the problem from a different point of view. There was a cultural perspective since the excessive use of plastic in the United States is linked to the consumption habit of single-use plastic bottles. There was a spatial perspective since human patterns in space interconnect with areas across the globe. Students acted and searched for a solution to reduce the consumption of single-use plastic bottles in our school community. After analyzing the invention of disposable water bottles, developing a risk/benefit chart, and determining the cost of different solutions; students designed a campaign to minimize the amount of trash that plastic bottles generate. As a result, our school community changed habits. They now bring their personal tumblers to the cafeteria to avoid using plastic water bottles.

Students needed to see the cause-and-effect factor. Plastics are useful, but the amount of plastic trash we generate accumulate and does not decompose. The implications on the environment and on our health can be catastrophic, so they took the necessary steps to address this community issue. Students sent letters to the principal, assistant principal, and the parent-teacher organization to request the installation of bottle-filling stations on campus, which were installed soon after. What a meaningful way to work together to generate an important change and create awareness about an issue in a positive way.

One of the most important advantages of culturally responsive teaching strategies is the value of building a relationship with students where they feel loved and accepted. Unconditional love is the foundation of all relationships and should be characterized by kindness, mutual respect, forgiveness, and empathy. Our students come from different backgrounds and experiences. Recognizing and validating our diverse students are the first steps to establish a relationship with them while we introduce different perspectives and relatable content to their personal experiences. We cannot have a relationship with someone we do not know. Getting to know our students builds a sense of belonging and identity that engages students in the truth that we are teaching in an environment where everyone is included.

STEM Blog ~ Notre Dame Center for STEM Education

The Notre Dame Center for STEM Education created a blog where I had the opportunity to share two articles. You will find many other articles from very talented Notre Dame STEM fellows that may fit better the content you teach and how you bring your personal identity into the classroom. These links will get you to the articles that were my contribution to the STEM blog. It is my hope that you continue exploring for ideas and inspiration in the blog.

Article "Food for Thought"

https://stemeducation.nd.edu/blog/blog-posts/food-for-thought

Article "Finding Authentic STEM Contexts in Our Own Backyard"

https://stemeducation.nd.edu/blog/blog-posts/finding-authentic-stem-contexts-in-our-own-backyard

Chapter 6

The Value of STEM in Food Science

Take care of your body, spirit, and mind because you are God's dwelling place.

Yanny Salom

We often do not think about the power and value of food, which keeps us healthy and supplies us with the energy we need to perform simple activities and life processes. There is so much STEM involved in those life sustaining compounds that are released when food breaks down. Scientists keep researching about chemicals in food, how they are processed in our body, how nutrients contribute to wellness, how yeast reacts in dough, what are the right amounts of each ingredient to produce food with nutritional value, how to understand labels, and so much more. By its nature, food science projects facilitate the integration of the students' life experiences by allowing connection surrounding self-awareness, health-related habits, and cultural associations.

More evidence about the value of STEM in food science and how it is linked to other areas is shown below:

- Agriculture: Farmers are using different methods to increase the quality and amount of food supply.
- Food safety: Providing the public with tips, news, and alerts on safely handling and storing food as well as monitoring the quality of food supply.
- Food processing: Improving methods to transform agricultural products into food. Food processing includes a variety of methods for preservation like curing, drying, canning, freezing, and controlled atmosphere; and food additives to improve the nutritional value, the flavor, the texture of food and to delay its spoilage.
- Researching: Learning about new ingredients and methods to provide food alternatives for people with alimentary restrictions.
- Creativity: Based on scientific principles, combining different ingredients to create delicious and nutritional dishes.
- Career connections: There are also career opportunities in the food industry that can help us feature the career connection as part of the STEM approach.

- Critical thinking: Analyzing and evaluating different ingredients and methods to create food choices as well as the interpretation of labels and information to have an evidence-based opinion on health claims for food products.
- Technology: How technology has made a contribution to the food we eat and the way we cook it from the discovery of fire to electricity, in addition to all of the new technologies in food preparation, storage, and processing.
- Global to local: Taking the global view of food to adapt it to our local community and personal needs. It also links with the availability of ethnic food, which is changing our way of cooking and eating.
- Geography: The food eaten is also linked to geography since it determines the type of crops that can be grown.
- The influence of other cultures in food customs: It has brought the practice of experimenting with foreign ingredients and methods, creating a fusion cuisine culture.
- Natural resources and their impact on food: How the quality of water, the nutrients of the soil, and climate affect crops.
- The role of selective breeding in food science: Methods developed by humans to produce tastier agricultural products with different textures and sizes as well as pest-resistant crops. This topic of selective breeding offers the opportunity to discuss how selective breeding can cause welfare and health problems when practiced on animals. The Catechism of the Catholic Church urges us to remember that "God entrusted animals to the stewardship of those whom he created in his own image. Hence it is legitimate to use animals for food and clothing. They may be domesticated to help man in his work and leisure. Medical and scientific experimentation on animals, if it remains within reasonable limits, is a morally acceptable practice since it contributes to caring for or serving human lives." (CCC # 2417).

The value of STEM in food science is limitless. One of my goals was to bring the food science curriculum to the classroom with a STEM approach. With the opening of the science lab, we implemented the Food and Nutrition program from the science curriculum *Full Option Science System* (FOSS)[29] where students performed investigations to measure the amounts of fat, sugar, and levels of acidity in various foods. The FOSS research-based curriculum allowed us to create awareness about the nutrients our bodies require and in which foods they can be found. To learn more about my insight about the value of STEM in food science, go to the Notre Dame STEM blog article "Food for

---

[29] To learn more about the FOSS curricula, go to https://fossnextgeneration.com/.

Thought."[30] We also brought programs from our local 4-H [31] organization, like teaching children how to make bread in a bag, that added value to our curriculum. 4-H is the largest youth development organization that offers a good starting point for programs on family and consumer sciences, among other things.

As stated previously, the value of STEM in Catholic education goes beyond the integration of disciplines, as our faith is the central element of this approach. To support this idea, topics on food science offer the opportunity to discuss our body as a God-given gift. As caretakers of our body, it is essential to understand that our body is a temple of the Holy Spirit. It implies that respecting our body includes accepting it, taking care of it, having a realistic image of it, and making sure that the language of our body speaks the truth. The book *Theology of the Body for Teens* states the idea that via body language our physical bodies reflect and allow us to see an invisible reality which includes our emotions. Educating the whole child includes teaching children that the human body reveals our character, our values, and that it is a gift from God which reveals our soul. This aids in establishing relationships based on mutual respect and love. To foster these values, health care and wellness professionals were always welcome.

Speakers were always a great addition to our STEM program. This was a way in which we shared information with our students about nutrition and exercise. The school counselors and nurses visited middle school classes to emphasize health concepts and encourage our students to develop good habits and lead healthy lives. Over the course of many years, our school nurse spoke about proper hand hygiene and the importance of consuming breakfast daily; with an emphasis on breakfast being the most important meal of the day. We created a Nutrition Newsletter which included reasons to eat breakfast and some ideas for quick, delicious, and healthy breakfast options.

A program including nutrition and exercise prescription for children and adolescents was introduced by speakers. Both of these concepts correspond to the reality expressed in the *Theology of the Body for Teens* of the body-soul composite; that the body reveals the soul, and we are our body and our soul as that is what makes us alive. A speaker discussed the importance of exercise during growth and development of the human person physically and emotionally as it can prevent chronic diseases if practiced regularly over the course of a lifetime. This lecture also touched on concepts from the *Theology of the Body for Teens*, emphasizing that since we are our bodies, then what we do with our body matters. Another speaker discussed food labeling, nutrition advice, and body image.

---

[30] Article "Food for Thought" https://stemeducation.nd.edu/blog/blog-posts/food-for-thought

[31] To find out more about your own local chapter, visit https://4-h.org/about/find/

One of the topics discussed was the unrealistic media images that may lead young people to eating disorders. Math and science were integrated to explain how weight and size are related and how the two have to correlate in order to have a healthy and realistic self-image[32].

Another food science topic we explored was hydroponic crop growth and the new technology being used. We built it with the cooperation of the Duval County 4-H organization where our main crops were lettuce and tomatoes. We celebrated the harvest with taco parties every year and parents could not believe their children were eating vegetables. Later, the school bought a hydroponics system for the STEM lab where we mainly grew peppers, basil, and kale. Once again, students were adding kale to their lunch meals and parents and teachers enjoyed the harvest in their own kitchens. Hydroponic farming is a sustainable method in a controlled environment with the additional advantage that it can be developed at home too. Since this approach does not require soil, growers provide the nutrients in the water, which allows plants to grow faster, fresher, and with the same nutritional level as vegetables that are grown from soil. Students eat the vegetables because they are consumed right after harvest, so they maintain properties of freshness, tastiness, and crunchiness. Growing food hydroponically alongside our embryology project were meaningful and impactful ways to teach our children agricultural skills. Both projects were done with the cooperation of our Duval County 4-H organization[33].

We went beyond Earth with our agricultural initiatives to help NASA with its goal to achieve long-term human presence in space. We took the challenge offered by the Growing Beyond Earth program related to food production in space. We followed a protocol to grow herbs and share experimental data online with NASA, after analyzing results and drawing conclusions. Each participant school receives a plant growth habitat comparable to the plant growing system at the International Space Station to recreate the conditions in space for this interdisciplinary science project.

Growing Beyond Earth is a project designed by Fairchild in partnership with NASA[34] where we were able to contribute data about the growth of basil. There were a variety of herbs in the investigation that were going to provide nutritional value and flavor to the prepackaged meals that astronauts consumed at the International Space Station, in addition to making the environment more pleasant for them.

---

[32] You can find meaningful information in the link below: https://www.chapman.edu/students/health-and-safety/psychological-counseling/_files/eating-disorder-files/13-barbie-facts.pdf

[33] To learn more about our local 4-H Duval County chapter, visit https://sfyl.ifas.ufl.edu

[34] To learn more about this initiative, visit https://science.nasa.gov/sciat-team/growing-beyond-earth/

Animals and food are two topics that will never fail to engage students. We created all kinds of physical models with candy, from the DNA molecule to cells in life science, to Bohr models of the periodic table of elements in physical science. Seventh grade classes always looked forward to showing their understanding of osmosis with gummy bears, while eighth grade classes looked forward to applying chemistry concepts to understand the science behind baking cookies.

Food processing is a very interesting field of food science too. As stated at the beginning of this section, it includes a variety of methods for preservation like curing, drying, canning, controlled atmosphere, and freezing, among others. One of the activities we did during STEM classes was the risk/benefit analysis of the canning process, which also led to the invention of can openers. Engineers are constantly using science to develop new technologies that solve problems. Many present-day cans have a ring pull to avoid the need for can openers. This design solved many safety concerns: the sharp edges of cans and dirty lids making contact with the product inside the can. Those risk/benefit analyses of inventions lead to new solutions or the improvement of existing solutions. In the analyses of inventions, I have found a meaningful way to promote critical thinking and class discussions while students become more critical about technologies and more responsible consumers of products and services.

To share an interesting project, the science behind baking cookies was one of my favorite projects to bring the chemistry component to a tangible and relevant level for students (see project description in the following pages). Forming students with a holistic approach includes understanding the academic component of chemical reactions and the takeaway of including cooking to develop societal and emotional connections as well as connecting recipes to various cultures, thus coming together to form unique personal experiences.

STEM Lab: The Science Behind Baking Cookies
Because Science Can Be So Sweet!

1. Watch the YouTube video "The Chemistry of Cookies" by Stephanie Warren. To watch the video, go to:

https://youtu.be/n6wpNhyreDE?si=IUbMOEAmCcNRosHO

We watched the video in class a couple of times. First, just to hear the information and then with a worksheet with questions (independent assignment). Students can watch the video as much as needed to complete the worksheet.

2. Independent Assignment: Building Understanding on the Science Behind Great Cooking and Baking. These questions are based on the information shared in the video" The Chemistry of Cookies" by Stephanie Warren.

- What are the three major steps of the baking process?
- Describe what happens when the cookie dough starts to heat up.
- At what temperature does the water in the dough turns into steam and the cookie starts to rise? Write the temperature value in degrees Fahrenheit and degrees Celsius.
- Which compound starts to break down into carbon dioxide?
- Name the two chemical reactions that fill the cookies with flavor and infuse the appealing brown appearance.
- Describe the process of caramelization.
- Describe the Millard reaction.
- What does the cookie's diameter depend on?
- What do you need to do to obtain a flatter, chewier, and wider cookie?
- What do you need to do to get a fluffier, cakier cookie?
- What do you need to do to obtain a thicker cookie? Explain the science behind it.
- How can you pump up the cookie's flavor and aroma? Explain the science behind it.

3. Class discussion of the independent assignment to clarify misconceptions and clarify concepts.

4. Comic Strip (small group activity): Use computer paper to create a comic strip that explains the science behind baking cookies. You must create a minimum of an eight-frame comic strip based on the video by Stephanie Warren. The comic strip needs to

illustrate the science behind baking cookies. You can choose a humorous approach, if it fits your group's idea to deliver the message, but it is not a requirement of this assignment.

5. Engineer the best cookie ever! (small group activity): As you learned from the video, you can play with the ingredients to tweak the chemical reactions and engineer the cookie of your dreams. According to the molecular biologist, Liz Roth-Johnson, "all this baking chemistry provides the building blocks for refining the cookie's architecture." You must type the following information: description of your cookie, ingredients, step by step directions, and an explanation to show understanding of the science behind baking cookies.

Giving each group the opportunity to bake the cookies in a kitchen setting would add great value to this learning opportunity. Keep in mind that it would require coordination between the school and the church, a short workshop on food and kitchen safety, and a group of parents who can assist the teacher in supervising the groups during the preparation and baking of the cookies. The book *Food for Today* by Helen Kowtaluk has a comprehensive chapter on kitchen basics (p.278-296) with information about food safety & storage and preventing kitchen accidents. Even though all these steps must be taken to include this part to the project, I believe it is worthy. It also represents a set of life skills that we must instill in our students.

Chapter 7

Creating and Coordinating an Astronomical Society

> "As I travel around, I tell people the answer is Jesus Christ, that Jesus walking on the earth is more important than a man walking on the moon."
>
> Astronaut James Erwin[35]

The founding of our Holy Family Astronomical Society (HoFAS) was one of my educational dreams for the students. I have a fascination for nature with an emphasis on the unknown universe and space in general. I soon discovered that many students shared the same fascination for Astronomy, the branch of science that deals with space and celestial objects. To emphasize a central Christian doctrine that is a distinctive sign of our faith, the quote stated above from Astronaut James Erwin provides the beautiful opportunity to reflect on the mystery of the Incarnation. Jesus coming to our planet as our Savior has given us the possibility of eternal life in Heaven, the goal which is at the heart of Catholic education. Educating the whole child calls for encouraging students to answer questions about the universe God has created through the lens of faith, and the astronomical society was the perfect space to continue wondering and loving our Creator.

The information below represents a piece of advice to establish an astronomical society:

- State your mission: define the purpose of the society, show what you are offering (broader goal).

Mission Statement of our Astronomical Society: to inspire and teach students to reach knowledge while looking beyond our Earth by reaching into space. This voyage, guided by science and technology, must always be accompanied by the responsibility to care for the universe God has created as a requirement of our faith.

---

[35] To read the article written by Clare Bruce about Irwin's Incredible Lunar Experience, visit https://hope1032.com.au/stories/faith/2019/jesus-on-earth-is-more-important-than-man-on-the-moon-the-legacy-of-astronaut-jim-irwin/.

Educating the Whole Child in Catholic Education

- Logo: visual representation of what you are offering/ identity of the society[36].

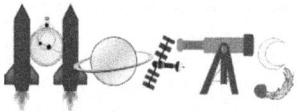

- Recruiting:
Aerospace Sign- Up table during school open house with:
  - Brochure of the astronomical society.
  - Letter to parents with mission statement, monthly activities, etiquette for stargazing nights, fees, and deadlines for joining the society.
  - Volunteers Sign Up Sheet.

- Promoting the Astronomical Society:
  - School Website
  - Church Bulletin
  - Social Platforms
  - Monthly email blast to parents/ School Daily Announcements

- Community Involvement:
  - Local 4-H Cooperative Extension Service: Organization that helps create and coordinate aerospace programs and offers activities on aerospace education.
  - Astronomical Societies (NEFAS is Northeast Florida Astronomical Society) that offer stargazing nights with powerful telescopes.
  - Universities (UNF and JU in Jacksonville): They usually have an astronomy club and offer stargazing nights (many of them open to public). They are also willing to serve as speakers.
  - Museums (MOSH ~ Museum of Science and History in Jacksonville): They have a planetarium with weekly events (mysteries of Mars, skies over Jacksonville, blackholes, etc.)
  - NASA: NASA.gov offers activities, competitions, videos, latest images in space, NASA TV live, among other things. To search for educational resources and opportunities, you may search by subject or grade level.
  - Parents/Volunteers: To get chaperones for field trips and volunteers for monthly activities, and as speakers or photographers.
  - National Coalition of Aviation Educators (NCAE)

---

[36] The HoFAS logo was created by Brooke Miller, former student.

- Air Force Association (AFA Florida Region for Jacksonville area): The organization has a STEM aerospace program in place, Civil Air Patrol (CAP), and STARBASE program (department of defense program that exposes youth to engaging STEM practices). STARBASE stands for Science and Technology Academies Reinforcing Basic Aviation and Space Exploration.
- Sally Ride EarthKAM program (**Earth K**nowledge **A**cquire by **M**iddle School students): Where students interface with a camera in the International Space Station (ISS) to take photographs of places on Earth. There are four missions a year and students have to review concepts on weather, latitude and longitude to request the images.
- Growing Beyond Earth[37] program: A citizen science program to help NASA on the initiative to grow plants in space. This program gives students the opportunity to be part of a controlled experiment that matches the conditions of the veggie habitat in the International Space Station (ISS) for growing food crops in space.

- Catholic identity
    - Include your Catholic identity in your mission statement.
    - Discussion with members on the quote by astronaut James Erwin "As I travel around, I tell people the answer is Jesus Christ, that Jesus walking on the earth is more important than a man walking on the moon." Other relevant quotes may also fit your specific needs and classroom culture.
    - Ways we show "responsibility to care for the universe God has created," which is part of our mission statement: Leading school to WWF adoptions as part of our Earth Day celebration, planting the butterfly garden to help monarch butterflies, joining recycling efforts within the school community, planting trees, creating videos about our planet with songs that are praising God's creation, among other initiatives.
    - Inviting the school pastor as a speaker to discuss science and faith topics.
- Recognizing members
    - End of the Year Celebration
    - Certificates and medals

---

[37] To learn more about this program, visit https://science.nasa.gov/sciat-team/growing-beyond-earth

- Aerospace Project Ideas
  - Aviation:
    - Attending an Air Show: Blue Angels
    - Building paper airplanes
    - Field trip to an airport
    - Demonstrating the principles of flight
    - Learning about aerial photography
    - Flying drones
    - Flying simulators
    - Identifying what/how flies: airplanes, helicopters, birds, butterflies….
    - Interviewing a pilot
    - Learning about people in aerospace
    - Learning how to become a pilot
    - Taking a flight lesson
    - Visiting a Military Air Base
  - Rockets:
    - Building rockets/ Rockets launch
    - Using a skyscope to measure the altitude and point of apogee of rockets
    - Exploring phases/parts of rockets
  - Space Exploration:
    - Designing a Space Station
    - Eating and exercising in Space
    - Exploring Space Travel History
    - Hydroponics in Space
    - Interviewing an astronaut
    - Moon Topography
    - Space Voyage Preparation
    - Visiting a Planetarium
    - Competitions on Space Exploration: Cassini scientist for a Day Essay Contest (participants wrote a 500 words proposal backing one of the targets for study for the Cassini Mission)
  - Careers in Aerospace:
    - Exploring Aerospace Careers
    - Learning about Civil Air Patrol and Reserve Officers' Training Corps (ROTC)
    - Interviewing/Shadowing people in aerospace jobs
    - Special thanks to the Air Force Association ~ Falcon Chapter # 399 for their outstanding Aerospace Education initiatives that support STEM education

in Northeast Florida. They offer a comprehensive project-based program that allows our youth to experience the wonders of aerospace and aviation through engaging hands-on activities.
- Weather:
  - Interviewing a meteorologist
  - Monitoring temperature, humidity, and pressure
  - Building/Reading barometers and thermometers
  - Field Trip to a TV Weather Station

Chapter 8

Teaching our Faith

> "For we walk by faith, not by sight."
>
> 2 Corinthians 5:7

We are not the product of random events. God has given us all a life with a purpose. He is continually moving us and provoking situations in order to fulfill his will that is always good and holy. It is our responsibility to live in that awareness to cooperate with God's plan. In that sense, I wore many hats as a teacher to fulfill God's will. I was hired to be the first full-time Spanish teacher; previously, parents volunteered to teach Spanish from the time that the school was founded. By the beginning of my second year as a teacher, I was given the opportunity to establish the science lab. The science lab was inaugurated in 2007, right after Christmas, and I started collaborating with teachers from kindergarten to sixth grade to validate scientific theories. I cannot thank the school and church leaders enough for the opportunity and for entrusting me to develop this initiative. The school continued growing, and one day God gave me the opportunity to teach religion to seventh graders along with my middle school science courses. He is a God of mercy and the gift of being one of the middle school religion teachers flooded me with many graces.

Educating the whole person within a Catholic setting has faith formation as a core component of the process. As such, it is the foundational piece from which all other subjects connect and integrate into. STEM education in Catholic schools follows the same principle, as religion is seamlessly integrated to see God in all things and to introduce the lens of faith within the academic content. To promote and support the interdisciplinary approach, reflecting on the lives of the saints as well as the affiliation of careers or activities was a relevant practice. For example, St. Albert the Great who is the patron of scientists and philosophers, Saint Hubert who is the patron of mathematicians and opticians, or Blessed Carlo Acutis who was an amateur website designer and programmer. Blessed Carlo Acutis was also a great model on how to use technology to serve God's Kingdom. He documented Marian apparitions and Eucharistic miracles on a website and became a reference on how to use technology to evangelize and interact with others. The responsible use of technology is connected with our faith and includes the proper and safe use of technology, which is an important skill to instill in students. Learning about scientists who were rooted in the Catholic faith was also a meaningful way to connect disciplines as we explored their convictions and scientific work. For example, Gregory Mendel was a Catholic priest who discovered the laws of inheritance or Nicolaus Copernicus who was a man of science and faith with great contributions to Astronomy.

The seventh-grade religion curriculum is delightful. It is not just the first-year preparation for Confirmation, but a journey that starts with the Infancy Narratives, continues with the Public Ministry of Jesus to close with the Glorification of the Lord (Resurrection, Ascension, Exaltation, and Pentecost.) The curriculum also includes the development of Gospels, the canonization process, Catholic Social Teaching, the Mystery of the Trinity, the sacraments, and the morality of a human act. To make it even more relevant, the *Theology of the Body for Teens* program is facilitated in seventh grade. The program was introduced to help children understand the importance of knowing our origin and destiny. This is presented as God's map to find our true identity with our origin as creatures created at the image and likeness of God and our destiny in Heaven. What a beautiful invitation for our children to see themselves as God sees them. The program truly encourages students to go beyond popularity, looks, academic performance, sports, and after-school activities to deepen their relationship with God, where their true identity lies.

Those years were enlightened with the Teaching Mass, a valuable initiative brought to our school culture by Father Timothy Cusick. He was also very generous with his time and visited the classroom to discuss relevant content on the Creation Story, evolution, biotechnology, morality, respecting life through the embryology project and as our astronomical society speaker. I also had a box where students wrote questions and comments about the curriculum we were discussing in religion class. The questions were sent to Father Cusick, and he visited the class for further clarifications. We also worked as a team to facilitate the chapter on vocation for the *Theology of the Body for Teens* program. Deacon Nullet, Father Cusick, and I were able to bring to the students three different examples of God's call and our responses with an emphasis that God's call is above all a call to holiness.

At a Catholic school, reality is both physical and spiritual, as we exist in an integrated universe that God has created. Pope Benedict XVI beautifully expressed that "the priest's mission is to combine, to link these two realities that appear to be so separate, that is, the world of God far from us, often unknown to the human being and our human world.[38]" When priests fulfill this mission, it also provides an opportunity for great conversations, establishing solid relationships, and sharing reflections on how we can live our faith. Students have questions and doubts that we should always welcome and respond to help them build understanding of our Catholic faith. Jesus's ministry has always been a source

---

[38] The "Lectio Divina of His Holiness Benedict XVI" (Hall of Blessings ~ Thursday, 18 February 2010)

of inspiration for me. When Thomas doubted about the Resurrection of Jesus, Jesus invited his disciple to touch his wounds. Jesus blessed those who believed without seeing him, but he also knows that doubts can lead us to increase and develop our faith.

The embryology project had students explore the life cycle of chicks. Students had the chance to observe how life develops by setting up fertile chicken eggs in incubators. They were responsible for the daily care of the eggs in the incubators until they hatched into newborn chicks. They were then responsible for the daily care of the hatchlings. This type of faith-based project deepens students' understanding of our perspective on the value and dignity of life. While students were enjoying the cuteness and beauty of the baby chicks, the simple but powerful statement "if you took care of the fertile eggs because there was life within them, then you should acknowledge a pregnant woman to be the carrier of a human life." What a great opportunity to remind students of the first theme of Catholic Social Teaching in which the Catholic Church proclaims that human life is sacred, and that abortion and euthanasia threaten the value of human life and the dignity with which God has created us.

Preparing students for their consecration to Jesus through Mary was another significant initiative. We used the book *Preparation for Total Consecration to Jesus through Mary* by Fr. Hugh Gillespie. This was a meaningful way to bring students closer to Jesus as we helped them recognize the role of Mary in the Mystery of Christ. Initiatives like these are tools for life. What students received from this experience cannot be measured through a summative assessment, but it represented a valuable resource for their spiritual formation and growth. St. Augustine stated, "You have made us for yourself, O Lord, and our heart is restless until it rests in you." (Chadwick 3). This quote is at the heart of Catholic education as it puts into words the sentiment that we are always aiming towards God.

The creation of the Mini Vinnies Chapter was another extracurricular initiative to address the Catholic Social Teaching themes known as *Option for the Poor and Vulnerable* and *Solidarity*. We need to put the needs of others first and love our neighbors, especially the poor. The Mini Vinnies organization is the youth chapter of the St. Vincent de Paul ministry where students engage in service projects, fundraisers, and pray for the most vulnerable in our society, the poor and the sick. This is a meaningful way to put our faith into action and lead by example as we instill the values of service towards others, empathy, and solidarity to build a more just world. The transmission of knowledge and culture to the following generation occurs when humanity understands the truth that every person is unique and unrepeatable; in doing so, we promote the value of life and the dignity of the human person as fundamental moral aspects in our society.

It is hard to choose one thing to share about the seventh-grade religion program. Creating brochures about Holy Week and designing Christmas cards according to the Infancy Narratives found in the Gospels of Matthew and Luke were some of the students' favorite memories. I also love the wonderful seasons of Advent and Lent that framed those projects, but going to Adoration and praying the Holy Rosary are two unforgettable memories. I cannot stress enough the value of praying with our children. However, I really learned so many lessons about how Jesus faced temptations; not just because we also face temptations, but because Jesus really showed us how to respond to temptations using the Word of God. Pope Francis shared a beautiful insight on how Jesus was led into the desert[39]: "the three temptations point to three paths that the world always offers, promising great success, three paths to mislead us: greed of possession- to have, have, have-, human vainglory, and the exploitation of God." Pope Francis added that Jesus shows us the remedies for temptations: "the interior life, faith in God, the certainty of his love – the certainty that God loves us, that he is the Father, and with this certainty we will overcome every temptation."

Knowing that temptations are daily challenges in our spiritual life, a "Temptation Survival Kit" activity (Campbell et. al. 93) was a meaningful way to encourage students to develop the mind of Christ and become more like him when facing temptations. The authors suggest having students decorate a shoebox and fill it with what helps them form their conscience and make good decisions. Some ideas of the contents include the Bible, prayers, sacramentals, and sports activities. Students then had the opportunity to share their survival kits and the reasons they chose each item with the class. To bring this important topic to a practical level, the article *Overcoming Temptation of Daily Life,*[40] portrayed five temptations and suggestions on how to become the people God created us to be.

The following chart shows how Jesus responded to temptations with the Word of God. This was a good strategy that allowed students to visualize and organize evidence of God's Words of wisdom from the Old Testament and the New Testament.

---

[39] Sunday Angelus - March 2019. To learn more, go to https://www.vatican.va/content/francesco/en/angelus/2019/documents/papa-francesco_angelus_20190310.html

[40] To read more about this article, visit https://www.loyolapress.com/catholic-resources/ignatian-spirituality/finding-god-in-all-things/overcoming-temptations-of-daily-life

Facing Temptation[41]

Temptation	Jesus's Response (Matthew 4, 1-11)	Deuteronomy
If you are the Son of God, command these stones to become loaves of bread.	One does not live by bread alone.	Dt. 8,3: in order to make you understand that one does not live by bread alone, but by every word that comes from the mouth of the Lord.
If you are the Son of God, throw yourself down; for it is written, He will command His angels concerning you, and on their hands, they will bear you up, so that you will not dash your foot against a stone.	Do not put the Lord your God to the test.	Dt. 6, 16: do not put the Lord your God to the test.
All these I will give you, if you will fall down and worship me.	Away with you, Satan! For it is written, worship the lord your God and serve only him. Then the devil left him, and suddenly angles came and waited on him.	Dt. 6, 13: the Lord your God you shall fear; him you shall serve, and by his name alone you shall swear.

As someone who immigrated from a foreign country, the lesson on the hardship in Jesus's life resonated with me. The Holy Family had to stay in Egypt until the death of King Herod, in the same way, many people across the globe have left their countries of origin in current times. Many of them, like Jesus, are refugees due to persecution, while others due to natural or man-made disasters, but in every case our response must be the practice of solidarity. Human suffering must be a call to action. When discussing Jesus's hardships and ministry, we must acknowledge that Jesus always favors the poor and the vulnerable. God's language is a language of mercy, empathy, and compassion to our brothers and sisters. This is the language we are called to speak as we honor the Holy Family as refugees and Jesus's ministry. Jesus advocated for the outcast, and he continues to do so as we work as his hands, feet, and voice in the world.

---

[41] To learn more about how to face temptations, visit https://www.loyolapress.com/catholic-resources/ignatian-spirituality/finding-god-in-all-things/overcoming-temptations-of-daily-life/

Chapter 9

An Integrated Approach for Foreign Language Acquisition

> "All of them were filled with the Holy Spirit and began to speak in other languages, as the Spirit gave them ability."
>
> **Acts 2:4**

I am inspired by the fact that at Pentecost, when the Holy Spirit spoke through the Apostles, the people listening were able to hear the Good News in their own language. Jesus brought the Gospel and united all people back then and continues to do so in the present moment. The study of a second language is aligned with the Catholic schools' mission of educating the whole person, as learning a foreign language opens our minds to new cultures, broadens our perspective, and deepens our respect for others. Language is vital for understanding others, building relationships, and spreading the Good News. Let us ask Jesus, the ever-present Teacher in the classroom, to unite us all in the reception of the Eucharist and to share the Gospel. Teaching Spanish was not only a rewarding experience, but it was also the way God led me back to teaching. When we immigrated to the United States, I thought my teaching career was over. Back then, I was able to read technical English in the science and engineering fields but writing and conversation skills were not present. Once I learned to speak English, I found it reasonable to teach Spanish. As a result, I decided to take a couple of courses to achieve my certification. At that point, I already believed that life-long learning requires relevant and interesting topics. Successful learning occurs when students are motivated, relaxed and confident. Both conditions are true for second-language acquisition and for any other discipline.

Based on my experience learning English as a second language, I knew students would better assimilate grammar content in Spanish, if they had already mastered it in English. Maintaining close communication with the middle school English teacher was crucial to success in introducing Spanish grammar content when students were knowledgeable of the equivalent content in English.

I also combined the Spanish curriculum with religion by teaching students how to participate in Mass celebrated in Spanish. We participated in the Spanish Mass at the diocesan level with eighth grade students, and had a school wide Spanish Mass. Learning about Marian apparitions in Spanish-speaking countries was also a meaningful connection that we combined with the Spanish Mass. The Latin American and the Caribbean people have a strong devotion to Virgin Mary. This devotion arrived with the

Spanish "conquistadores" who evangelized indigenous people, resulting in their conversion to the Roman Catholic faith. We also established the routine of praying the Hail Mary and the Our Father in Spanish class as a way to foster our Catholic identity.

Integrating Spanish with the Social Studies curriculum was also relevant and appropriate. Lessons relevant to the geography of the Spanish-speaking countries became common lessons for both subject areas. We also discussed exploration and celebrated the arrival of Christopher Columbus in the Americas to commemorate the encounter of distinct cultures. The history of Florida was also part of the curriculum as the arrival of Ponce de Leon marked an important part of our local history with important implications for our Catholic faith including the founding of the city of St. Augustine, which became the center for Catholic missionaries in the southeastern region of North America. Other co-curricular lessons occurred by integrating art and music to the curriculum. We learned children's songs in Spanish to build vocabulary and we designed piñatas in cooperation with the art teacher. Art is part of the culture of countries, and we discussed many artistic manifestations in order to expose students to the culture of the Latin American countries.

"Middle school learners establish a connection between language learning and their real lives and interests in order to be motivated to learn." (Shrum and Glisan 129) In that sense, we practiced the vocabulary learned by role-playing a visit to the doctor or pretending we were registering for the Olympic games. The idea was to offer relevant and interesting activities in the target language, so students were not thinking they were interacting in another language. These types of initiatives also complied with the characteristics of effective programs for middle-school learners which, according to the same authors, must develop the students' ability to communicate effectively in real-life situations.

"Design of curricular activities around communication goals and activities to develop skills to shop, eat in restaurants, and plan trips to develop communication skills" (Hall 59) are other effective ways to develop skills in the target language. One of the activities we enjoyed was bringing different toys to create a supermarket where students had to buy and pay for products. However, the most popular activity was our annual field trip to a Hispanic restaurant where students had to order and communicate in Spanish during the field trip. To ensure a successful experience, students received the menu in advance, and we role-played ordering in a restaurant during class. After the field trip, students went back to the restaurant with their families and ordered in Spanish for the entire family. Planning a seven-day trip to a Spanish-speaking country was a meaningful project too. Students chose a country of their interest and planned the entire trip: plane tickets, sites, itineraries, shows, possible restaurants, and recommendations for traveling. It also included a paragraph about why they had chosen that country, the description of two sites of interest, and writing a postcard to their Spanish teacher.

"Some activities that are appropriate for beginning and intermediate students where topics that elicit personal information and simple statements about preferences, concrete experiences, and the like" (Omaggio 298). One of the activities proposed is pretending that students who are on vacation write a postcard. In the "plan a trip" project, students had to write a postcard to explain where they were, what they were doing, what they liked about the place, who was with them, and any other additional information they considered relevant. The author of this book also suggested writing a Christmas card for the teacher. We also did an activity where students had to wish a merry Christmas, share their plans for the Christmas break, and find a Biblical verse in Spanish that was appropriate for the season. Both assignments, the postcard and the Christmas card, had to be written in Spanish in paragraph format.

Taking into consideration the self-image perception of middle-school students is vital as describing the physical appearance of self or others may be uncomfortable. The description of pets, vacations, rooms, pictures, and artwork in general is a safer approach. The use of appropriate folktales, among other things, to offer engaging activities was also effective. We implemented the folktale "The Three Little Pigs" in Spanish. This is a known folktale, which facilitates understanding and engagement. Students were divided in groups of five (narrator, three pigs and the wolf) and they presented the story with the proper characterization (costumes) and using Spanish vocabulary including cerditos, lobo, paja, bosque, ladrillo, madera, soplar, derrumbar, tengo hambre, tengo miedo, tengo dolor, among others. "An extension activity to the folktale initiative, like talking about favorite parts, speculating on why an event occurred, and expressing opinion about a character" (Shrum and Glisan 195), was relevant to the folktale project. I usually added the task of writing a different ending to practice writing skills. Exposing students to legends, cartoons, poems, and songs was also relevant, as well as the use of recipes to practice communication skills and learn more about the Spanish culture.

"The use of real or simulated travel documents, hotel registration forms, biographical data sheets, train and plane schedules, authentic restaurant menus, labels, signs, newspapers, and magazines" (Omaggio p.97), was an effective way to use authentic language in instruction. Other ideas from this author that I implemented with certain modifications, like providing prompts, for teaching writing at the lower level were:

- Identify objects in a room with a picture supplied by the teacher or allowing students to create their own picture or diorama of a chosen room.

- Ask students to create or complete an incomplete menu for a restaurant they are planning to open. The description of the dishes must be included.
- Express likes and dislikes.

Teaching the Hispanic culture is essential to encourage students to get engaged and motivated to learn the target language. Teaching students how to cook and taste some foods, as well as cultural manifestations such as dancing, painting, games, and songs are authentic ways to immerse students in the culture.

Even though having a strong understanding of the content you are teaching is crucial, the development of teaching methods is also decisive to successfully deliver knowledge. I have a deep respect for the field of education and everyone who wants to teach must develop a set of skills and tools to facilitate instruction in an effective way. As stated before, the acquisition of knowledge happens when we make instruction relevant, accessible, interesting, and fun. In addition to that, we must create a positive learning environment where students feel supported, relaxed, motivated, encouraged, and confident that we can reach them exactly where they are.

Miscellaneous

Words of Advice ~ Graduation Dinner ~ Class of 2022-2023

HoFAS Article from AFA Falcon Newsletter – Spring Issue 2017

Butterfly Garden

STEM Lab Article

Blessed Carlo Acutis STEM Laboratory and Plaque Statement

Words of Advice
Graduation Dinner ~ Class of 2022-2023

It is with great joy that I accepted this opportunity to speak to you before your graduation. When Mr. Moloney asked me to give you a piece of advice at your graduation dinner, I asked him to allow me to pray about it. I went to Adoration that day and it became clear that Jesus was allowing me the opportunity to speak words of encouragement he has for you. The following quote is attributed to the Catholic writer and apologist, Gilbert Kieth Chesterton:

"Right is right if nobody does it; wrong is wrong even if everybody is wrong about it."
This may sound familiar to you. It echoes the clear and simple truth: that right is right and wrong is wrong. Keep this in your heart and mind every time you have to make a decision.

I know this day and age are challenging. Social platforms and social media have brought upon us a falsehood of "virtual reality," which is not reality at all. It has created all kind of pressures and misconceptions about who you truly are. Keep in mind that who you are to God, is who you truly are. He imprinted his image and likeness in each one of you and he created you for greatness. Believe in your goodness and remember that quote I kept on my board for the last two quarters "whenever you feel unloved, unimportant, or insecure, remember to whom you belong." This is one of my favorite quotes because it represents a great reminder that God's presence and abiding love runs far deeper than any brokenness, pain, or hurt that we may experience in our earthly journey. This is also very clear in Isaiah 43:1 "Fear not, for I have redeemed you; I have called you by name, you are mine."

I have experienced so much joy in the classroom watching how good you can be! I have also exercised patience when your choices do not align with the person God created you to be. God created us in his image and gave us the gift of freewill. Love, in order to be true must be free and this gift comes with great responsibilities. The *Catechism of the Catholic Church*, section 1733, advises us that "the more one does what is good, the freer one becomes. There is no true freedom except in the service of what is good and just". Believe in your goodness and practice it as you exercise this freedom.

When people asked me from where I get my energy, joy, and inspiration despite the personal and professional challenges, my answer is always "I keep my eyes fixed on Christ." Whenever you feel discouraged, frustrated, sad, happy, or amid any other emotion; keep your eyes on the Lord.

Last but not least, stretch your heart to create and receive holy moments. Become God's coworker as you become the hands, feet, and voice of Christ. Make yourself available to God and his people and do what you prayerfully believe God is calling you to do. Give yourself a fresh start and believe in God's mercy because "the rest of your life begins the moment you make yourself available to God." (Kelly 37) I will add one more reason; because the voice, hands, and feet of Christ belong to those willing to be moved by God. Let us then close this holy moment as we have begun each science and STEM class: "Blessed be Jesus. Blessed be his Holy Name." To which you always responded: "Now and forever more."

Forever in my prayers and in my heart my dear and good children of the light. Thank you for all the holy moments we have created together.

Jesus, Mary and Joseph; Holy Family, pray for us.

HoFAS Article from AFA Falcon Newsletter

**AFA FALCON CHAPTER
NORTH EAST FLORIDA**

Astronomical Society (HoFAS) hosted a stargazing night

By Yanny Salom, M.S., Falcon TOY for 2015
Holy Family Catholic School

The Holy Family Astronomical Society (HoFAS) hosted a stargazing night recently. About 30 members and their parents joined us and were able to take a closer look of some planets: Saturn and its magnificent ring system; the red and the second smallest planet in the solar system, Mars; the farthest known planet in the solar system, Neptune; and a planet that rotates in the opposite direction to most other planets, Venus. We also observed the Vega Star, the brightest star in the constellation of Lyra and the second brightest star in the northern hemisphere (Arcturus is the brightest). HoFAS members also observed the binary star Albireo in the constellation Cygnus, also known as Beta Cygni. This is a very unique double star with one blue star and the other golden. The Ring Nebula was also visible during our stargazing night. Its shape is very distinctive and is used in many astronomy books, but new observations by NASA reveals that the nebula's shape is more like a jelly doughnut. It ties perfectly with our snack for the stargazing night, galactic doughnuts!! The Ring Nebula is located in the constellation Lyra and is considered a popular target for amateur astronomers like our HoFAS members. Eighth grade students will start a project on constellations soon so our stargazing night was a great way to motivate our students before the project on constellations.

All of these was possible due to the kindness and love for astronomy of the Northeast Florida Astronomical Society (NEFAS). We thank NEFAS for bringing this learning experience to our HoFAS members this year again. Non HoFAS members also joined us and we were really excited to share with the school community as well. The moon did not come out until midnight last Friday but it did not stop us from enjoying a wonderful night and other wonderful celestial bodies. The Heavens are telling the glory of God and His wonderful creation.

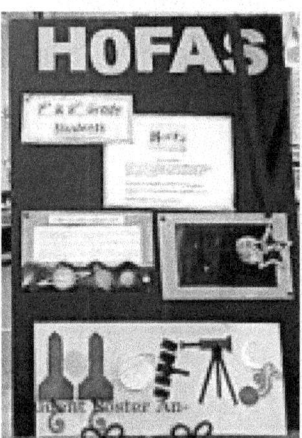

A Student Poster Announcing the event posted as a reminder

Yanny Salom M.S., Science Teacher is verifying the students observation findings.

A student observing stars & constellation with assistance from astronomer Mr. Greg Sauve, NEFAS President

Prepping one of the star gazing telescope is Mr Greg Sauve, President Northeast FL Astronomical Society

- - -A Note on What is NEFAS? - - -

NEFAS, or the Northeast Florida Astronomical Society, is an organization of amateur astronomers who live, serve and observe in Northeast Florida. They operate primarily in Jacksonville, Florida and its surrounding communities. NEFAS members typically give "sky tours" utilizing green laser pointers to indicate objects in the sky. This is a great way to become familiar with the night sky for all ages. A public observing session is an evening where NEFAS members set up their telescopes and invite the public at large to come do some astronomical observing. Members can answer questions, give advice on telescopes to purchase, and point out interesting objects in the night sky. Public observing sessions are geared toward all ages, and can be enjoyed by adults as much as by young children.

Butterfly Garden

Having a butterfly garden has given students the opportunity to see the life cycle of butterflies in our school backyard, but this is also the place where we release butterflies that we grow in the lab thorough the 4-H program "Becoming a Butterfly." 4H is America's largest youth development organization that provides free resources and projects for students to develop academic and life skills. Some other activities that we have developed in our butterfly garden are:

1. Using conceptual models to classify the plants and living organisms in the garden. Seventh grade students created charts for each organism with the eight levels of classification of living things and the scientific name of each living organism.
2. Building an equity garden where each student from PK to 8th grade created their own design in a river rock. We placed all the rocks in a special bed with a message about how we are all part of the Mystical Body of Christ.
3. Using it as a memorial garden.

Butterfly Garden Layout

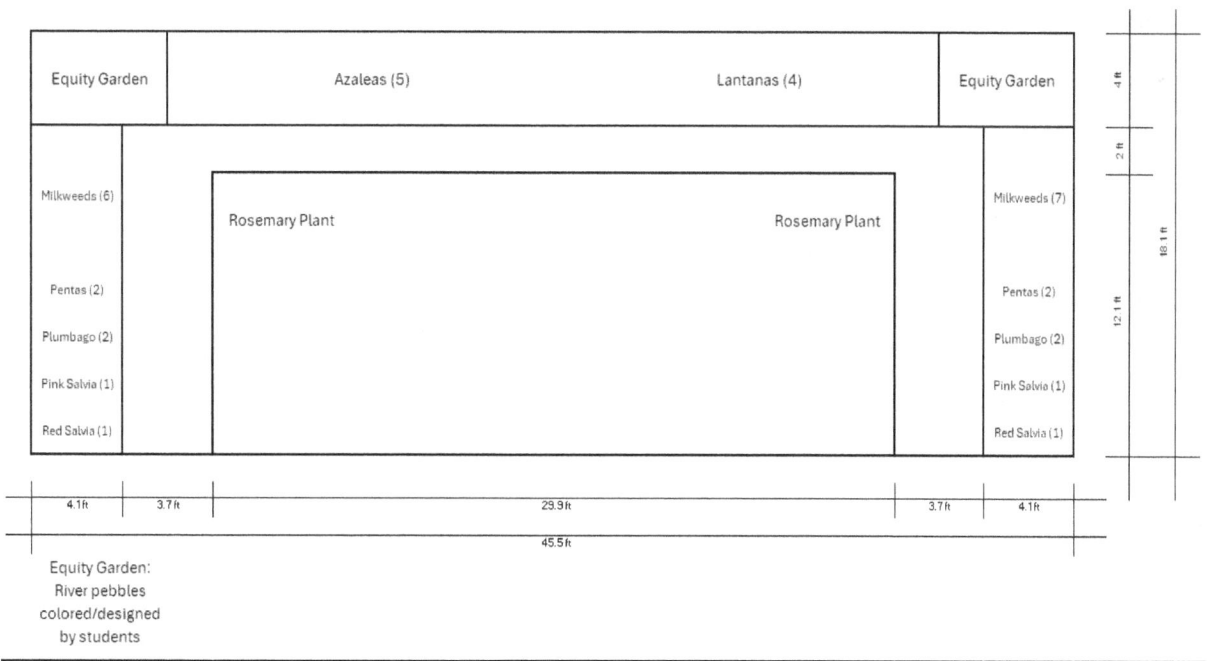

Our New STEM Lab (August 2022)

Holy Family Catholic School is celebrating its 20th anniversary, and we are thrilled to mark this milestone by welcoming students into the new, state-of-the-art STEM lab this school year! The STEM lab initiative will follow the pillars of the school's mission statement. Students will be exposed to an integrated approach of disciplines with curriculum alignment and real-world problem solving that will expand students' access and engagement, deepen their understanding, and create career awareness.

Providing resources to support STEM integrations throughout the curriculum is an essential element of our program. One of the many new resources to enhance the K-8 STEM program is the Maker Space work area where students will collaborate, explore, and create with simple and complex technology like with our 3D printer. Using Maker Space, students can do many things including creating 3D models and electronics to coding. The coding area will have a section for elementary students with Bee-Bots while middle school will use Sphero Robot Balls. The Chibitronics© technology will provide an additional layer for coding as students learn about electrical circuits. Our Maker Space movement began with projects on sustainable energy and will continue adding more areas that you can check out at https://teachergeek.com/collections

Just as "academic excellence" is an important pillar of our school mission statement, so too is "concern for others." As stewards of all God has created, we will continue our program on protecting the environment. We strongly believe and live out our motto, "conservation is the foundation of environmental protection." In that sense, we continue positioning our students as helpful humans who are part of the solutions to the world's environmental challenges. We will continue taking care of the habitats for the monarch butterflies and the Eastern bluebirds we have on our school property as we add more initiatives that will promote the right mindset for our students to value God's creation. We will start a hydroponics garden in the STEM lab, which will empower students to understand the value of fresh food, where food comes from, and the different nutrients necessary for plants to grow. This initiative, along with the embryology and butterfly projects we do with the Duval County 4-H extension, will teach our students farm-homemaking practices. To learn more about 4-H, a youth development program that is part of the University of Florida, go to https://sfyl.ifas.ufl.edu/duval/4-h-youth-development/. Another pillar of our school mission statement is "responsibility for self." Students will explore food science and sports science. Students will learn to understand the value of a healthy lifestyle that includes making educated food choices and the importance of exercise from scientific points of view. Most importantly, we will continue to guide students to "foster their faith." The essential STEM practice embraced in our new lab includes the integration of academic disciplines within our Catholic faith as our worldview, which reflects our existence as individuals and as a school community. Jesus Christ is our ever-present model as he is the integrating principle of everything. May we all follow the steps of Jesus and become his hands, feet, and voice for the glory of the Father. Thank you to the parents, parish and community members who helped fund and support the STEM lab and Holy Family Catholic School throughout these first 20 years. The investment you have made in our students is a powerful gift toward their future successes to become coworkers with God.

Blessed Carlo Acutis STEM Laboratory[42]

Blessed Carlo Acutis is a modern-day role model for the next generation of Catholic students and future saints. Below are some of his quotes and how they translate into important Catholic values.

Blessed Carlo Acutis's Quote	What the Quote Affirms
"By standing before the Eucharistic Christ, we become holy."	This affirms the teaching that Jesus unites us all in the reception of the Eucharist, calling us to the vocation of holiness.
"All people are born as individuals, but many die as photocopies."	This affirms the truth that the human person is single, unique, and unrepeatable. He speaks to the importance of being who God created you to be and the freedom that comes from living in the truth of who you are. The truth that living a whole-hearted and integrated life brings to light the reality of evangelizing with our whole being at all times.
"To always be close to Jesus, that's my life plan."	The life of Blessed Carlo Acutis is a great Christian witness and proclamation of the Gospel Truth to all students that step foot in the STEM Laboratory. Jesus is the integrating principle of everything.

Plaque Statement

The essential STEM practice embraced in our new lab includes the integration of academic disciplines within our Catholic faith as our worldview, which reflects our existence as individuals and as a school community. Jesus Christ is our ever-present model as He is the integrating principle of everything.

---

[42] To learn more about Blessed Carlo Acutis's quotes, visit https://www.ncregister.com/blog/17-inspiring-quotes-from-carlo-acutis.

Works Cited

"17th Century Maritime Celestial Navigation." YouTube, youtu.be/5DO1wygRtPY. Accessed 5 May 2024.

Air Force Association Falcon Chapter #399 Inc, Spring 2017, Astronomical Society (HoFAS) hosted a stargazing night, Salom 2017, www.falcon.afaflorida.org/Newsletters/FalconSpring17.pdf

"Angelus, 10 March 2019, 1st Sunday of Lent: Francis." Angelus, 10 March 2019, 1st Sunday of Lent | Francis, 10 Mar. 2019, www.vatican.va/content/francesco/en/angelus/2019/documents/papa-francesco_angelus_20190310.html.

Austriaco, Nicanor, "How Does God Create Through Evolution?" *Thomistic Evolution*, 2020, www.thomisticevolution.org/wp-content/uploads/sites/182/2020/05/Thomistic-Evolution-23.pdf.

Austriaco, Nicanor, "The Fittingness of Evolutionary Creation." *Thomistic Evolution*, 2020, www.thomisticevolution.org/wp-content/uploads/sites/182/2020/05/Thomistic-Evolution-22.pdf.

"A Century of Screams: The History of the Roller Coaster." PBS, Public Broadcasting Service, www.pbs.org/wgbh/americanexperience/features/coney-century-screams/. Accessed 24 Jan. 2023.

"Bioethics Materials." *USCCB*, www.usccb.org/prolife/bioethics-materials. Accessed 4 Mar. 2023.

Blue, Laura, and Sarah Redick. "Science for all: Socially transformative teaching." *Science Scope*, vol. 45, no. 4, 2022, pp. 48–55, https://doi.org/10.1080/08872376.2022.12291470.

Brent, James, "The Nature of Creation." *Thomistic Evolution*, May 2020, www.thomisticevolution.org/wp-content/uploads/sites/182/2020/05/Thomistic-Evolution-7.pdf.

Bruce, Clare. "An Encounter with God on the Moon: Astronaut Jim Irwin's Incredible Lunar Experience - Hope 103.2." *An Encounter with God on The Moon: Astronaut Jim Irwin's Incredible Lunar Experience*, 19 July 2019, hope1032.com.au/stories/faith/2019/jesus-on-earth-is-more-important-than-man-on-the-moon-the-legacy-of-astronaut-jim-irwin/.

Butler, Brian, et al., *Theology of the Body for Teens: Leader's Guide*. Ascension Press, 2011.

Campbell, Barbara F. and Campbell James P., *Finding God Following Jesus*. Loyola Press, 2014

Cartier, Jennifer L., et al., *5 Practices for Orchestrating Productive Task-Based Discussions in Science*. NCTM, National Council of Teachers of Mathematics u.a., 2013.

Catechism of the Catholic Church. United States Catholic Conference, 1994.

Chadwick, Henry, *St. Augustine Confessions*. Oxford University Press, 2008.

"Curricula." *EiE*, https://yes.mos.org/curricula/. Accessed 12 Jan. 2024.

Davenport, Thomas, "God's Knowledge and Love in Creation." *Thomistic Evolution*, May 2020, www.thomisticevolution.org/wp-content/uploads/sites/182/2020/05/Thomistic-Evolution-8.pdf.

Dispezio, Michael A., et al., *Science Fusion – Cells and Heredity*. Holt McDougal/Houghton Mifflin Harcourt, 2012.

Dispezio, Michael A. and Marjorie Frank, et al., *Science Fusion – Matter and Energy*. Holt McDougal/Houghton Mifflin Harcourt, 2012.

Dispezio, Michael A. and Michael R. Heithaus, et al., *HMH Science – Dimensions. Engineering & Science*. Houghton Mifflin Harcourt, 2018.

"Fides et Ratio (14 September 1998): John Paul II." *Fides et Ratio (14 September 1998) | John Paul II*, 14 Sept. 1998, www.vatican.va/content/john-paul-ii/en/encyclicals/documents/hf_jp-ii_enc_14091998_fides-et-ratio.html.

"Find Your Local 4-H." *National 4-H Council*, 18 Jan. 2024, 4-h.org/about/find/.

"Foss Next Generation - NGSS Certified Science Curriculum." *FOSS*, 1 June 2023, fossnextgeneration.com/.

"FOSS®." FOSS, foss.lawrencehallofscience.org/. Accessed 4 Jan. 2021.

Franciscus. *Laudato Si' on Care of Our Common Home*. Our Sunday Visitor Pub, 2015.

"Get Real Barbie" Fact Sheet*, www.chapman.edu/students/health-and-safety/psychological-counseling/_files/eating-disorder-files/13-barbie-facts.pdf. Accessed 16 Aug. 2023.

Gillespie, Hugh, and Louis-Marie Grignion de Montfort. Preparation for Total Consecration to Jesus Christ through Mary: According to St. Louis De Montfort. Montfort Publications, 2013.

"Growing beyond Earth - NASA Science." NASA, NASA, science.nasa.gov/sciact-team/growing-beyond-earth/. Accessed 22 Feb. 2022.

Hall, Joan Kelly. *Methods for Teaching Foreign Languages: Creating a Community of Learners in the Classroom*. Merrill, 2001.

"Helping Sea Turtles Survive since 1959." Sea Turtle Conservancy, conserveturtles.org/. Accessed 7 Mar. 2023.

Kelly, Matthew. *Holy Moments: A Handbook for the Rest of your Life.* Blue Sparrow, 2022.

Kloser, Matthew, and Sophia Grathwol. *Reading Nature: Engaging Biology Students with Evidence from the Living World*. National Science Teaching Association, 2018.

Kowtaluk, Helen. *Food for Today. Student Ed.* Ninth ed., Glencoe/McGraw-Hill, 2006.

Launius, Carrie and Christine Anne Royce, "What Makes a Good STEM Trade Book?" *NSTA/Blog*, NSTA, 13 Jan. 2020, https://www.nsta.org/blog/what-makes-good-stem-trade-book. Accessed 9 Nov. 2023.

"Let Your Creativity Shine with Paper Circuits." Chibitronics, chibitronics.com/. Accessed 30 June 2022.

"Meeting with the Parish Priests of the Diocese of Rome: Lectio Divina (February 18, 2010): Benedict XVI." Meeting with the Parish Priests of the Diocese of Rome: Lectio Divina (February 18, 2010) | BENEDICT XVI, 17 Feb. 2010, www.vatican.va/content/benedict-xvi/en/speeches/2010/february/documents/hf_ben-xvi_spe_20100218_parroci-roma.html.

Murphy, Alyssa. "First Millenial Saint: 17 Inspiring Quotes from Soon-to-Be-Saint Carlo Acutis." *NCR*, 10 Oct. 2020, www.ncregister.com/blog/17-inspiring-quotes-from-carlo-acutis.

"Next Generation Science Standards." Home Page | Next Generation Science Standards, www.nextgenscience.org/. Accessed 9 June 2023.

Omaggio Hadley, Alice. *Teaching Language in Context*. Third ed., Heinle & Heinle, 2000.

"Overcoming Temptations of Daily Life." Loyola Press, 15 Jan. 2025, www.loyolapress.com/catholic-resources/ignatian-spirituality/finding-god-in-all-things/overcoming-temptations-of-daily-life/.

"Professional Development." National Geographic Society, 2 July 2024, www.nationalgeographic.org/society/education-resources/professional-development/.

"Resources in Catholic Bioethics." *The National Catholic Bioethics Center*, www.ncbcenter.org/bioethics-resources. Accessed 4 Mar. 2023.

"Saint Bernardine of Siena, and the Power of the Name of Jesus: The Holy Name of Jesus." *The Holy Name of Jesus Saint Bernardine of Siena and the Power of the Name of Jesus Comments*, www.holyname.ie/saint-bernardine-of-siena-and-the-power-of-the-name-of-jesus/. Accessed 8 May 2023.

Salom, Yanny, "Finding Authentic STEM Contexts in Our Own Backyard." *Notre Dame - Center for STEM Education*, Notre Dame, 7 Feb. 2019, https://stemeducation.nd.edu/blog/blog-posts/finding-authentic-stem-contexts-in-our-own-backyard. Accessed 11 Jan. 2024.

Salom, Yanny, "Food for Thought." *Notre Dame - Center for STEM Education*, Notre Dame, 17 Sept. 2020, https://stemeducation.nd.edu/blog/blog-posts/food-for-thought. Accessed 9 Jan. 2024.

"Sea Turtles." SeaWorld, seaworld.com/educational-resources/sea-turtles/. Accessed 12 Jan. 2023.

Shrum, Judith L. and Eileen W. Glisan, Teacher's *Handbook: Contextualized Language Instruction*. Third ed., Thomson, Heinle, 2005.

Stemeduc. "Center for STEM Education." Center for STEM Education, stemeducation.nd.edu/. Accessed 23 June 2022.

"Symbolic Species Adoptions." *WWF Gifts*, World Wildlife Fund, gifts.worldwildlife.org/gift-center/gifts/Species-Adoptions. Accessed 2 Oct. 2023.

The Catholic Youth Bible: New Revised Standard Version, Catholic Edition. Saint Mary's Press, 2010.

"The Chemistry of Cookies - Stephanie Warren." *YouTube*, YouTube, 19 Nov. 2013, www.youtube.com/watch?v=n6wpNhyreDE.

Thomistic Evolution, www.thomisticevolution.org/. Accessed 2 June 2024.

Tinkercad, www.tinkercad.com/. Accessed 15 Mar. 2022.

Towey, Jim, *To Love and Be Loved: A Personal Portrait of Mother Teresa*. Simon & Schuster, 2022.

Trustey Family STEM Teaching Fellows Notebook

UF/IFAS Extension – University of Florida, sfyl.ifas.ufl.edu/duval/4-h-youth-development/. Accessed 4 Nov. 2023.

Vasquez, Jo Anne, et al., *Stem Lesson Essentials, Grades 3-8: Integrating Science, Technology, Engineering, and Mathematics*. Heinemann, 2013.

Vost, Kevin. "Albert the Great - Light of Science, Light of Reason." *Catholic Answers*, 2023, pp. 43–45.